Contents

Wild, Wild World of Animals

Animal
Defenses

A TIME-LIFE TELEVISION BOOK

Editor: Charles Osborne
Associate Editors: Bonnie Johnson, Joan Chambers
  Author: Ogden Tanner
  Writers: Cecilia Waters, Deborah Heineman
  Literary Research: Ellen Schachter
  Text Research: Maureen Duffy Benziger, Judith Gies
Picture Editor: Judith Greene
  Permissions and Production: Cecilia Waters
Designer: Robert Clive
  Art Assistant: Carl Van Brunt
Copy Editor: Eleanore W. Karsten
Copy Staff: Robert Braine, Florence Tarlow

WILD, WILD WORLD OF ANIMALS
TELEVISION PROGRAM
Producers: Jonathan Donald and Lothar Wolff
This Time-Life Television Book is published by Time-Life Films, Inc.
Bruce L. Paisner, *President*
J. Nicoll Durrie, *Vice President*

THE AUTHOR

OGDEN TANNER, a former senior editor for TIME-LIFE BOOKS, writes on nature and other subjects. In addition to having written articles on coyotes and on the impact of man on the environment, he is the author of books in the TIME-LIFE BOOKS American Wilderness series and of *Bears & Other Carnivores* and *Beavers & Other Pond Dwellers* in the Wild, Wild World of Animals series. A native New Yorker and an architecture graduate of Princeton University, he has been a feature writer for the San Francisco *Chronicle*, associate editor of *House & Home* and assistant managing editor of *Architectural Forum*.

THE CONSULTANTS

COLIN BEER is Professor of Psychology at Rutgers University. He has taught also at Otago University, New Zealand, where he was an undergraduate, and at Oxford University, where he obtained his doctorate. He does research on the communication behavior of gulls and has published numerous articles on this subject and on more general aspects of animal behavior.

ERIC QUINTER graduated from Pennsylvania State University in 1969 with a Bachelor of Science degree. He is a curatorial assistant at The American Museum of Natural History in New York City and, under a grant from the National Science Foundation, is revising a monograph on noctuid moths.

DR. RICHARD G. ZWEIFEL is Chairman and Curator in the Department of Herpetology of The American Museum of Natural History in New York. His fields of study include the ecology and systematics of reptiles and amphibians, in particular those of America and New Guinea. Dr. Zweifel has published more than 70 scientific papers in addition to articles for magazines and encyclopedias. His memberships include the American Society of Ichthyologists and Herpetologists, the Herpetologists League, the Society for the Study of Amphibians and Reptiles, the Ecological Society of America and the Society for the Study of Evolution.

COVER: Treehoppers, disguised as spiny green thorns, plunge their tubular proboscises into a slender stem and ingest the sap. Each insect automatically aligns itself with its fellows to face in the same direction.

Wild, Wild World of Animals

Animal
Defenses

Based on the television series
Wild, Wild World of Animals

Published by
TIME-LIFE FILMS

The excerpt from Pilgrim at Tinker Creek by Annie Dillard, © 1974 by Annie Dillard, is reprinted by permission of Harper & Row, Publishers, Inc., and Blanche Gregory, Inc.

The excerpt from "How the Leopard Got His Spots" from Just So Stories by Rudyard Kipling is reprinted by permission of The National Trust and The Macmillan Company of London and Basingstoke, courtesy of A. P. Watt, Ltd.

The excerpt from Wild Season by Allan M. Eckert, © 1967 by Allan M. Eckert, is reprinted by permission of the author.

The excerpt from Jim, the Story of a Backwoods Police Dog by Charles G. D. Roberts, © 1929 by Macmillan & Company, Inc., is reprinted courtesy of the publisher.

The excerpt from Season on the Plain by Franklin Russell, © 1974 by Franklin Russell, is reprinted by permission of the author.

ISBN 0-913948-23-3.

Library of Congress Catalogue Card Number 78-61808.

Printed in the United States of America.

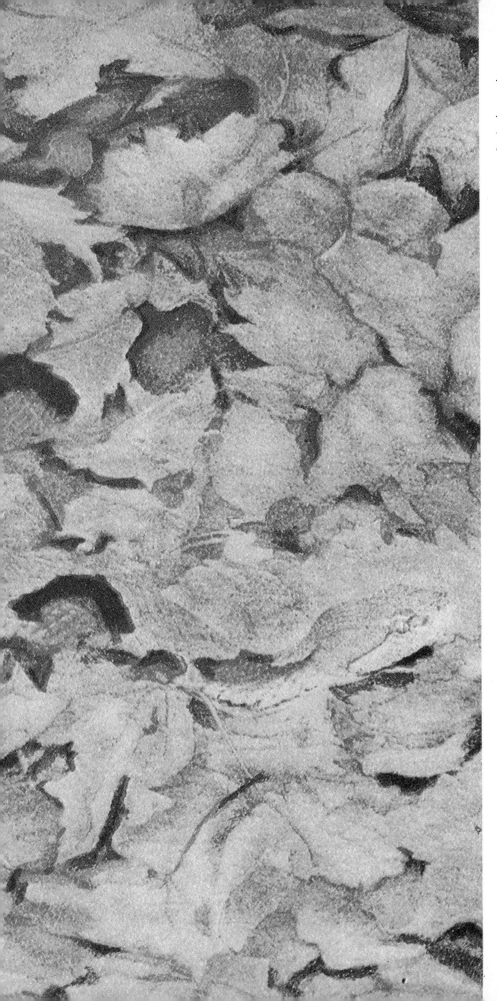

Introduction

by Ogden Tanner

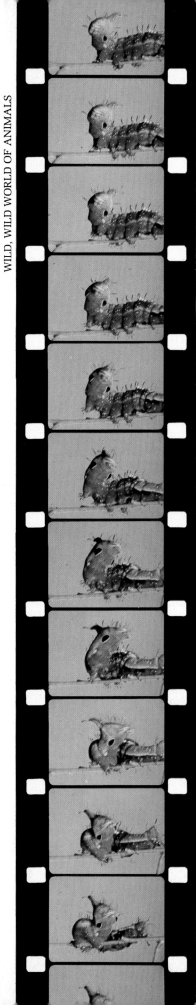

IN THEIR AGE-OLD STRUGGLE for survival, animals have developed an astounding array of protective adaptations designed for a single purpose: to avoid being killed by other animals, including man. Generally, these attributes are divided into two categories called primary and secondary defenses. Primary defenses are those mechanisms that operate whether or not an animal is in the presence of a predator. For example, creatures like moles and earthworms simply avoid attack by living their entire lives underground, in a hole or hidden away in some retreat—a way of life called anachoresis. Other animals have bodies that are colored, patterned or shaped in such a way as to make them virtually indistinguishable from their surroundings, either because they blend in with their background—a phenomenon called crypsis—or because they are disguised to resemble an inanimate, inedible object in their habitat, such as a twig or a thorn. Some animals with dangerous or unpleasant attributes, such as the foul-tasting ladybug or the toxic poison-dart frog, make no secret of their powers and actually advertise their potential to predators by means of their vividly marked bodies or by other signals such as a strong smell or a distinctive sound. This stratagem, called aposematism, is an extremely effective defense technique. Numerous non-noxious species also mimic the appearance of these unpalatable, highly visible animals—with the result that predators steer clear of them too.

Not all brightly colored animals are aposematic, however. Some use their striking coloration as an attraction during courtship or as aggressive displays in territorial disputes with members of their own species. This volume, however, will deal only with interactions between predators and prey—including the fact that some predators employ coloration and other adaptations usually associated with defense in order to hunt and kill more effectively. The bold stripes of the tiger, for instance, are a form of camouflage called disruptive coloration; the markings help the big cat conceal itself in the tall Indian grasses as it stalks its prey.

In contrast to primary defenses, secondary defenses are brought to bear when an animal actually encounters a predator. Some animals, when pursued, try to frighten a predator away by means of bluffs or threats. This response, called deimatic behavior, is typified by birds that ruffle their feathers and cats that raise the hair along their backs in order to appear larger and more formidable than they really are. Other animals live in large social groups such as herds and flocks, and the members of these groups work cooperatively to fend off attackers. The most common secondary defense mechanism, however, is flight. Given the opportunity, most animals will choose to flee before resorting to the final option: to fight.

Defensive attributes are frequently the most notable characteristics of many animals, contributing in large part to their shapes, colors and behavior—as well as to man's primary images of them. The deer is well known for its big eyes and ears and its long, graceful legs: The sensory organs are designed to alert the deer to danger, and the limbs are superbly suited to getting the animal safely away at the highest possible speed. The black-and-white stripes of the

African bark spider

skunk constitute a visual warning, alerting men and other animals to its celebrated weapon, much as the distinctive black-and-yellow bands and the constant buzzing of a bee advertise its potent sting. Some species rely on sheer size and/or ferocity to keep other animals at a respectful distance: Few lesser beasts care to tangle with a full-grown elephant or rhinoceros, a tiger, an eagle or a wolf.

Whether or not they possess such obvious attributes, most animals employ a variety of stratagems, calling into play successive lines of defense in the event that primary or even secondary ones do not work.

Consider, for example, the eastern hog-nosed snake. A harmless, non-poisonous species, it depends initially, like many other vulnerable creatures, on a passive defense: Aided by the camouflage provided by its natural body markings, the hognose freezes the moment it senses danger so that no movement will catch the eye of an attacker and give the snake away.

If found and confronted in spite of such defenses, the hognose shifts to stratagem No. 2. It tries to appear larger and more frightening and powerful than it really is, flattening its neck in the intimidating manner of the poisonous cobra, hissing loudly and flicking its tail menacingly back and forth. Often this display is convincing enough to startle an enemy into hesitating before it attacks, allowing the snake a precious instant in which to employ stratagem No. 3: slipping quickly away.

Should the threat fail to fool the invader, the hognose, like the opossum and

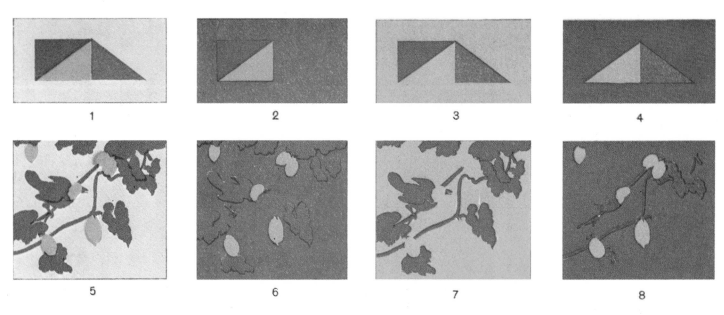

These two diagrammatic sequences illustrate a basic principle of defensive camouflage called disruptive coloration. The drawings show how each of the color patches on an animal's body is obliterated when seen against a background of the same color; the effect is to blur the animal's body outline. Diagrams 1 through 4 make the point using a tricolored geometric shape; diagrams 5 through 8 show a comparably colored bird. When viewed against a white background, all three colors are visible and the figure is easily discerned. But when placed against a background matching any one of its colors, the figure's silhouette is drastically altered. When this principle is realized in nature, it reduces a predator's ability to perceive its prey's body outline.

a few other animals, switches to stratagem No. 4: playing dead. It abruptly flips over on its back, mouth agape, and perhaps fakes a convulsion or two to add a realistic touch before lying still as death. This unexpected development generally baffles a predator, particularly one that instinctively goes after only live, moving animals in order to eat; the predator usually decides to break off the engagement, abandoning a motionless but highly edible meal. So dedicated an actor is the hog-nosed snake that if a human observer turns it right side up with a stick, the snake will promptly flop over on its back again as though to prove that it really is dead—peeking occasionally to see if the enemy has left so that it can revert to the safer stratagem of slithering away, out of the predator's reach.

In heavily populated and highly competitive habitats it seems that the smaller the creature the more complex its adaptations; and the defensive techniques of animals, in the words of *Alice in Wonderland*, get curiouser and curiouser. In Central and South America, for example, insects of the praying-mantis family perch motionless in trees and bushes, their bodies so perfectly specialized to resemble twigs, leaves or flowers that even sharp-eyed birds have difficulty spotting them. Some moths that are active mainly at night have outer wings almost indistinguishable from the bark of the particular trees they rest on during the day; if they are detected in spite of their camouflaging, they abruptly reveal markings on their inner wings that look like huge hawk or owl eyes. These can surprise a smaller predatory bird and allow the moth a split second to get away. Other moths have developed false heads, eyes and antennae at their rear ends in order to divert attacks to body parts that are less vital than heads and visual organs. Still others have perfected high-pitched clicks that they can emit to "jam" the sonar-like apparatus that hunting bats use to locate them.

Even more astonishing are the evasive tactics employed by such deep-sea denizens as squid and cuttlefish, which can change colors almost instantly to match the changing complexion of the sea bottom over which they pass. As a backup strategy, they can also release a smoke screen of opaque fluid should an invader get too close.

One of the most remarkable items in the grab bag of strange animal defenses is that of the tiny bombardier beetle, which is barely larger than the ants that sometimes assault it in hopes of a meal. If the attacker has not learned by experience to heed its flamboyant warning colors—bright blue and orange— the bombardier mixes an instant cocktail for its hapless guest by squeezing chemicals from storage tanks in its body into an activating enzyme chamber located at the tip of its abdomen. The result is an immediate explosion that drives a smokelike burst of irritants, heated to the boiling point, right into the attacker's face.

How did such marvels of adaptation come about? This question has long intrigued biologists and animal behaviorists, who have barely begun to find the answers. With their limited intelligence, animals could hardly have "invented" their defense and weapons systems in the sense that man has dis-

In escaping an attack by a predator, a pursued prey animal often follows an erratic course to prevent the predator from predicting its next move. The illustration shows the circuitous escape route taken by a moth that is being hunted by a bat. The moth evades its predator by unexpectedly veering upward just as the bat closes in for the kill.

When zebras are hunted by packs of hyenas, they respond defensively by forming family groups in which the foals are surrounded by mares and yearlings. The stallion falls to the rear, occasionally attacking the pursuers. Hyenas keep up a long chase, hoping to tire some zebras and force them to lag behind where they can be brought down.

guised his own weaknesses with an awesome arsenal ranging from rifles and machine guns to nuclear bombs. A mantis could not possibly have created its uncanny resemblance to a dead leaf as a feat of magic or a conscious act of will. How, then, could that resemblance have come about?

The key, as the great 19th-century naturalist Charles Darwin discovered, lies in the vast and fantastic process called natural selection, which has involved myriad variations among uncounted billions of animals over millions of years. An ancient ancestor of the dead-leaf mantis, for example, must have been born with a random combination of genes that made it look slightly more like a dead leaf than its fellows did—perhaps with a brown-tinged spot on a generally green body or the suggestion of leaflike veins along its wings. As a result, when it rested among dead leaves it was slightly harder for insect-eating birds and other predators to detect; for this reason it and similarly patterned members of the species survived in slightly greater numbers than the ordinary members of the tribe. In turn, they produced more offspring that closely resembled themselves. As fewer leaflike members of the population were eaten, they became more numerous, and the "dead leaf" characteristic became

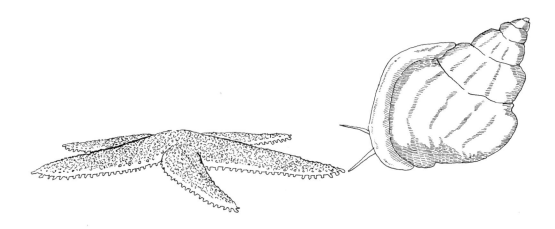

more pronounced and widespread. At the same time, however, the predators of the mantis were also evolving. In order to survive, they were becoming more adept at spotting flaws in their prey's disguise. The mantis population in turn developed even more convincing traits, including ragged-looking wing edges and spots that looked like holes that had been chewed by caterpillars, as well as a swaying motion to imitate the movement the real leaves surrounding them made in a breeze.

Similarly, a wide range of evolutionary changes must have taken place in all the other species of animals in existence today, perfecting in each its peculiar method or methods of defense in response to specific predators in specific habitats. The startling false eyes, or eyespots, of certain moths—which also occur in caterpillars, fish and other creatures—may well have begun as simple, dotlike imperfections that were just different enough to cause an attacker to hesitate, thus giving their owner an advantage; gradually, as predators concentrated on moths that lacked the dots, these random spots became more prevalent in the survivors, and evolved into more and more efficient—and frightening—replicas of real eyes. (In some species the eyespots have bands of

The defense mechanism evolved by gastropods, or snails, specifically to be used against predatory starfish is shown above. The snail responds to the starfish's touch (above, left) by extending and flexing its foot—and jumps out of the starfish's grasp (above, right). By twisting its shell to face in a different direction, the snail is able to crawl away.

13

color that precisely mimic pupil, iris and even reflected highlights; at least one type of moth, by contracting its wing muscles, can actually make the "eyes" appear to blink.)

Effective as any given defense may have become through natural selection, however, the animal is under constant pressure to change it; animals are continually evolving ways of overcoming their prey by means of keener senses, stealth, speed or other special techniques. The mongoose has learned how to avoid the poisonous fangs of the cobra and to kill through sheer agility and fighting skill; bears and badgers have developed an immunity to beestings as they root for honey in a hive. To overcome the defenses of shrimp, which burrow in the sand, the cuttlefish sweeps the sea bottom with a jet of water to uncover its prey; as its target gives away its position by hastily trying to cover up again, the cuttlefish shoots out its tentacles to grab the meal. Even the tough, close-fitting shells of clams and oysters are no guarantee of protection: Starfish can exert a pull strong and long enough to pry them open, while the oyster drill, a snail, uses its specially developed tongue to bore a feeding hole through the shell and suck out the contents.

Some biologists have likened such prey-predator relationships to an endless arms race, in which the prey species are constantly evolving more efficient methods of defense while the predators are constantly evolving more efficient methods of counteracting them. The result is a kind of escalating equilibrium, with natural enemies competing on more and more sophisticated levels all the time. Sometimes a species fails to adapt to the race or to other changes in its environment, and becomes extinct. Among the rest, imperfect individuals are continually being culled; only the fittest, healthiest, best-adapted members on both sides of the equation survive to reproduce.

Man, who represents only one of the millions of species that has evolved over the eons, reveals his heritage in many ways. Being an animal himself, he often behaves like more primitive species when he is in danger. Faced with a crisis, he may freeze in his tracks or cringe in an instinctive attempt to make himself appear smaller and less noticeable. Or he may obey the basic animal impulse to run, or attempt to hide or camouflage himself by melting into the crowd. If unable to avoid a direct confrontation, he too reacts with bluffs or warnings, drawing himself up to his full height, clenching his fists in a threatening gesture, even baring his teeth as many animals do when they are cornered and see no way out.

Man also makes use of more refined stratagems of animal defense. Soldiers' uniforms are drably colored or mottled and military installations camouflaged with protective patterns to make them more difficult for an enemy to detect. Many fortifications bear remarkable resemblances to the shelters that animals construct against attack: As beavers dam ponds and build their lodges in the center, men have built castles surrounded by moats; the bombproof underground complexes of recent wars have much in common with the intricate passages, living quarters and emergency exits of an ant colony or a prairie-dog town. To shield their vulnerable bodies, warriors through the ages have de-

Before the Industrial Revolution, the peppered moth was a light-colored insect (above). Its speckled coloration helped camouflage the moth against its usual background—similarly patterned, lichen-covered trees. Gradually, however, the peppered moth evolved a dark pigmentation (below) to blend with soot-blackened tree trunks—the grimy legacy of England's industrial progress.

vised armor and shields that imitate the plates and scales of animals such as armadillos, and armored cars and tanks that resemble the shells of turtles. The wagon train drawn up in a circle against Indian attack exemplifies the same strategy of group defense as that used by herd animals like musk oxen, in which the females and the young huddle in the center of the group while the fighting bulls, with pawing hooves and lowered horns, form a circle around them.

As a technological animal, man has had a disproportionate impact on other animals' survival and defenses. By relentless hunting and eradication of natural habitats, the human race has shown that it can push other species over the edge of extinction. Yet there is evidence that, given time, animals can adapt to the changing conditions of a modern world. Many species have become warier as a result of persecution; some, like coyotes, are intelligent enough to do most of their wandering near human settlements at night while the inhabitants sleep—and have adapted their diets opportunistically to include the contents of garbage cans. Some crop-destroying insects have responded to pesticides by developing mutant strains that are immune.

Perhaps the most provocative example to date of man's impact on evolution is the case of *Biston betularia*, the peppered moth, which along with other moths in Europe and North America has exhibited an evolutionary wrinkle scientists call industrial melanism. Before 1850 a good 99 percent of the peppered moths in Britain, as their name suggests, were of a grayish-white color speckled with black, superbly camouflaged for their daytime resting places on the lichen-covered trunks of trees; only a few mutants were of a highly visible all-black, or melanic, form. With the advent of the Industrial Revolution, however, coal particles and fumes from factories in Manchester and other cities began to kill off the lichens on the trees and to leave a sooty black deposit in their place. Against such a uniformly dark background, the typical peppered moth stood out clearly and thus became an easy target for birds; now it was the black variety that enjoyed the advantage of camouflage. Under the pressure of natural selection the situation gradually reversed. By the turn of the century 95 percent or more of the peppered moths around Manchester were black, while those in unpolluted rural areas elsewhere in Britain remained predominantly grayish white.

Since the 1950s, researchers studying the phenomenon more closely have made two more discoveries. One was that the gene structures of the melanic moths had apparently altered to capitalize on the growing grime; captured specimens were actually found to be darker than those first observed more than a hundred years ago. The second discovery was still more interesting. Though the black form still prevails, its numbers are beginning to dwindle. In recent years, positive results from regulations against air pollution indicate that if man can influence evolution he can also reverse his impact on the environment and even restore things to their preindustrial state. As a result of cleaner air, the lichens are coming back to the trees of Manchester, and so is the original coloration of the peppered moth.

Cover and Concealment

One of the commonest defenses among animals is retreat into cover or concealment—under or behind some sort of barrier in order to make detection difficult or to discourage attack. These hiding techniques are as varied as is the animal kingdom itself. Some species spend all, or almost all, of their lives hidden from possible attack. This phenomenon—called anachoresis, after the Greek word meaning to retire or withdraw from the world—can be observed not only in such familiar creatures as moles and earthworms but also in burrowing lizards, amphibians and mollusks that dig themselves into sand or mud, as well as in many insects that bore tunnels beneath the bark of trees.

A few anachoretes have so perfected their techniques that they can excavate permanent and virtually impregnable homes in wood and rock where no other creatures can penetrate. An especially intriguing—and destructive—example is that of the shipworm, which despite its name and appearance is not a worm but a bivalve mollusk related to the clam. It lives only in wood that is immersed in sea water. Using the rasplike projections on its shells to scrape away the wood, which provides nourishment as well as protection, the shipworm digs a burrow about a quarter of an inch in diameter and up to a foot long. The burrows of thousands upon thousands of shipworms can riddle the hull of a boat or the wooden pilings of a dock; since they are undetectable from the surface, they may go unnoticed until one day, without warning, the dock may suddenly crumple into the water or the boat may break up and founder.

More numerous are animals that live in holes or nests only part time, and those that take cover mainly in emergencies. Many small mammals—rabbits, wood-chucks, badgers, voles—dig dens in which they hide from predators during the day, emerging toward nightfall, when they are not as easily seen, to forage for food. Denning by day is particularly common among desert creatures, which do their hiding and sleeping in cool underground quarters not only to avoid predators but to escape the blazing, desiccating heat. Many birds on the other hand roost by night in nests or holes they have prepared in order to hide themselves and their young from nocturnal hunters. They come out by day when they can use their keen eyesight to spot insects and when they can easily see potential enemies and make use of their major defense of flight.

In their evolution of defenses against predators many burrowing and nest-building animals have developed special techniques; examples are dummy nests, multiple burrows, false entrances and emergency exits. Marsh wrens may build as many as a dozen fake nests to reduce the chances that a fox or raccoon will discover the real one, which is carefully concealed. Some mammals, notably coyotes, also inhabit several dens, moving their pups from one to another if they sense that the first one has been discovered. Beavers and muskrats build their lodges in ponds, entering and leaving them by underwater passages that their less adept predators cannot reach. The lesser swallow-tailed swift constructs a tubelike nest that is suspended from an overhanging rock face or a tree. The bird enters the nest through an opening in the bottom, but to deflect intruders it places in the side another, more conspicuous entrance, which ends in a blind alley.

Among the more fascinating of the hiding animals are those whose own body architecture is designed as a kind of portable refuge. The best known are the turtles, whose ability to withdraw into their tough shells has ensured their survival for 175 million years. Other species rely on different kinds of armor. Alligators and some lizards are covered with horny projections and plates; fish like the sturgeon and gar are sheathed in scales so hard and close set that they can deflect a small-caliber bullet fired at close range. Some armadillos and pangolins, whose upper bodies are clothed in jointed plates, roll up to protect their soft undersides, presenting a tight ball almost impossible for an attacker to pry open.

Various creatures depend on shell-like armor to protect them from attack. Clams, oysters, mussels and scallops retreat into their twin shells and close them tightly; snails and conches pull their soft bodies into single shells and block the openings with horny plates attached to their feet. Crustaceans like lobsters and crabs, as well as insects, have segmented bodies enclosed in an external skeleton strengthened by a tough, flexible material called chitin. The hermit crab, lacking the normal armor of other crabs except at its front end, appropriates the empty shells of other sea creatures like snails. It tucks its vulnerable body into the shell, using its large right claw to block the entrance when forced to retreat. Since it regularly outgrows its borrowed shell, the hermit seems to be constantly house hunting, examining every new shell it comes across, fighting with and sometimes evicting another crab in order to establish itself in a new mobile home.

Toad burrowing in Sahara

The Diggers

Going underground is the primary defense of many species of animals. By hiding in a burrow or covering up with mud or sand an animal can conceal itself from both predators and prey. But burrowing is also important at times when an animal cannot easily defend itself, such as when it is sleeping or caring for its young. Weasels, for example, nurture their defenseless young in the relative safety of an enclosed nest built either above- or belowground. And ghost crabs take to their burrows during the molting process, called ecdysis—when they have shed their old shells and are waiting for their new ones to harden.

Other animals use subterranean retreats when conditions aboveground become unbearable: too dry, too hot or too cold. The side-winding adder and Eastern spadefoot toad escape desiccation by going underground, where it is often 35° cooler than it is on the surface. Northern animals burrow through the snow for similar reasons—it can be 50° warmer under the snow than above it.

An Eastern spadefoot toad pauses at the entrance to its burrow. To avoid dehydration during a drought, these animals may stay underground for months until adequate rain falls.

A short-tailed weasel peers out from its burrow. These short-legged carnivores often appropriate the homes of mice and ground squirrels—first eating the original inhabitants.

A side-winding adder (above) can sink
under the sand in seconds, leaving its
tail exposed as a lure.

Sheltered by the door of its burrow, a
trap-door spider (below) emerges just
enough to catch unwary insects.

The Harpa (above), a species of mollusk, uses its muscular foot to dig into the mud (below). When it is submerged, the Harpa retracts its body and foot into its shell for extra insurance against predators.

A ghost crab (right) digs downward with sweeping motions of its limbs. Its waterline burrow may extend a yard below the surface of the sand, but must often be rebuilt after inundation by storms or high tides.

Having a Ball

When attacking another animal, most predators aim for the most vulnerable parts of their victims' bodies—usually the abdomen or the head. To protect themselves from such an attack, such passive animals as the pangolin, chiton and millipede have evolved a covering of horny plates or scales that extends along the length of their backs and on their heads. If threatened, these animals can roll into compact balls, thereby shielding their soft, vulnerable underparts and making it almost impossible for an attacker to distinguish one end from the other.

The chiton, a primitive mollusk, differs from other species of mollusks, which have either a single or hinged shell into which they retreat when threatened. In sharp contrast, the chiton has eight flat hinged plates that bend around its body for protection—earning the name "coat-of-mail shell." Millipedes have similar sets of protective armor, but in addition they have certain secondary defenses that are deployed to discourage persistent predators. Some of these millipedes secrete lethal cyanide; others emit an alkaloid that tastes bitterer than quinine.

The pangolin is one of the largest of these ball-rolling animals—often growing six feet long and weighing as much as 60 pounds. Both the terrestrial and the arboreal species sleep in this defensive position, either in a ground den, in a tree cavity or wedged in the fork of a tree. To make sure her young are protected while she dozes, a mother pangolin nestles them close to her underside and then wraps herself around them.

Having rolled itself into a near-perfect sphere, a pill millipede (left) is virtually impenetrable for a predator.

A pangolin (above) begins to wrap its long, thick armored tail around itself to protect its scaleless underside.

A flattened chiton (below) adheres to an underwater surface. If dislodged, it rolls into a ball for safety.

A heteropod sticks its proboscis out of its protective shell (above). This tiny mollusk eats plankton by ripping it apart with its abrasive tongue.

In the sequence below, a transparent Phronima crawls into the case of a salp (below, left), eats the occupant, and takes over the shell (below, right).

Just like its larger mollusk relatives, a pteropod (right) retracts into its shell for safety. To swim, it extends its foot and flaps it back and forth.

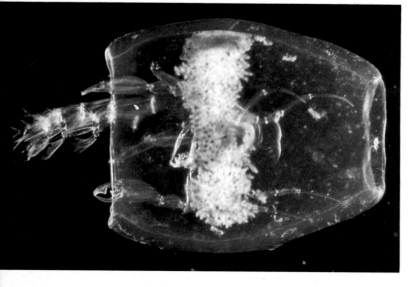

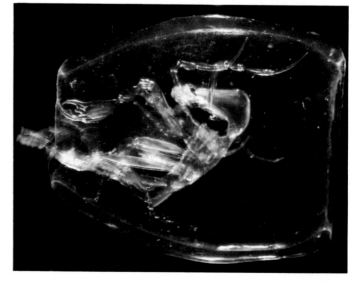

Tiny Shelters

Animals that hide or bury themselves underground for protection are guaranteed safety only if they can get to the burrow or cover themselves before a predator can attack. The solution to this problem—especially for a slow-moving creature like a mollusk or a turtle—is for the animal to carry its own shelter in the form of a shell or case.

Small mollusks like heteropods and pteropods come in both snail- and bivalve-like forms. Some heteropods even have an operculum—a hard plate on the bottom of the foot that is used to plug the shell's entrance. Animals without their own coverings often take over those of other creatures. The tiny *Phronima sedentaria* appropriates the case of a salp, a relative of the sea squirt, which provides a safe retreat for the *Phronima* and its young.

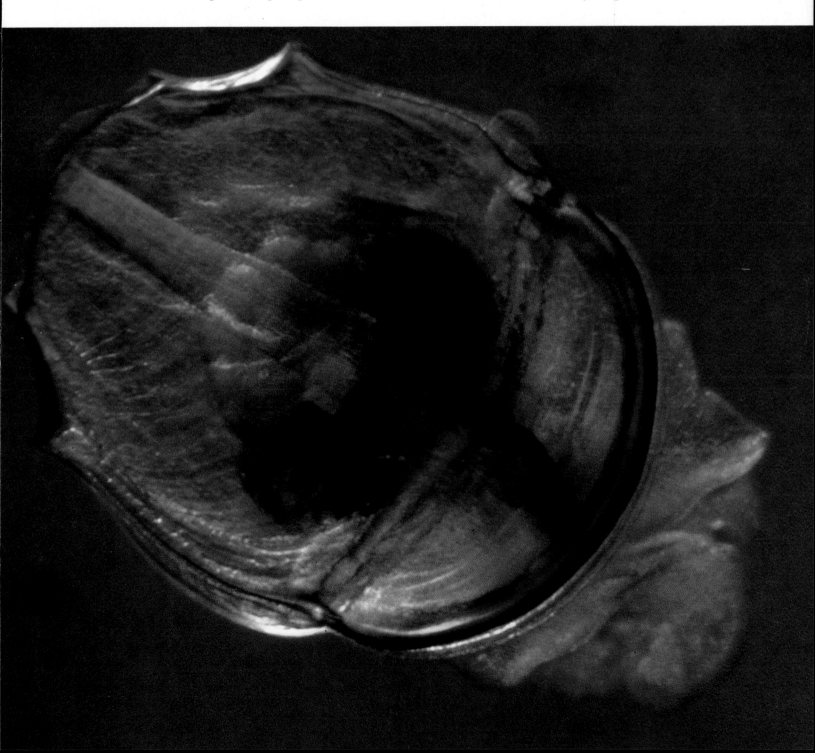

Armored Abodes

If any single trait characterizes all 200 species in the order of turtles, it is the shell. A highly effective adaptation that has characterized turtles and their ancestors, in one form or another, for millions of years, the shell has enabled the turtle to adjust to both terrestrial and aquatic habitats. Turtles use their shells to retreat from danger—drawing their soft, vulnerable bodies into impenetrable armor-plated fortresses.

The shells have two sections. Both the top section, or carapace, and the bottom part, or plastron, have two layers: Typically, the outer one is composed of horny scales, the inner of tightly jointed bones. The carapace and plastron are usually joined at the sides by a bony bridge, with openings at the front and back for the animal's head, legs and tail. In some species, such as the three-toed box turtle, the plastron is hinged in the middle and can be folded in front and back to seal these openings completely.

The turtle's shell grows along with its occupant, and if it is damaged, the carapace has remarkable powers of regeneration. Some fishermen still hunt the endangered tropical hawksbill turtle, dip its back into boiling water, and then peel off the turtle's horny carapace plates, which they sell as valuable tortoiseshell. If the animal is then turned loose, and if it is young enough, it is thought to be able to regenerate the lost plates.

A three-toed box turtle (left) peeks out from between its shells. Box turtles can fold their hinged plastrons, or lower shells, sealing themselves inside.

A leopard tortoise (above) bears its massive-looking shell with ease. To find food these animals must often undertake long migrations.

A Communal Life

Some of the smallest and most defenseless animals in the world erect extremely successful protective structures by cooperating with other members of the same species to build communal homes. A termite community, for example, begins when a breeding pair digs a small chamber just a few inches below the ground, where they mate and the female lays her eggs. These offspring become the first workers to start construction on the aboveground termite castle. With the aid of succeeding generations of workers, the castle can become a massive structure that may eventually house well over a million individuals.

Certain species of ants have developed a different kind of communal home—residing within the ball-shaped bases of the thorns of an acacia tree. The tree provides food and shelter. In turn, the ants protect the tree by stinging any browsers that try to eat their living room.

Two ants crawl toward the entrance of their acacia-thorn home. Ants eat the pulp inside the thorn's base and then live in the cavity they have created.

A termite nest (right) can be more than 20 feet tall. Such an edifice, with its series of spires and network of tunnels within, may take years to construct.

Pilgrim at Tinker Creek

by Annie Dillard

Annie Dillard was awarded a Pulitzer Prize in 1975 for Pilgrim at Tinker Creek. *The book records a year's observations of the wildlife in the vicinity of Dillard's house in the Roanoke Valley of Virginia. In this excerpt she describes and reflects on the life that is hidden under the surface of the earth. She is a contributing editor for* Harper's *magazine and has also published a volume of poetry,* Tickets for a Prayer Wheel, *and a work of spiritual exploration,* Holy the Firm.

Earthworms in staggering processions lurch through the grit underfoot, gobbling downed leaves and spewing forth castings by the ton. Moles mine intricate tunnels in networks; there are often so many of these mole tunnels here by the creek that when I walk, every step is a letdown. A mole is almost entirely loose inside its skin, and enormously mighty. If you can catch a mole, it will, in addition to biting you memorably, leap from your hand in a single convulsive contraction and be gone as soon as you have it. You are never really able to see it; you only feel its surge and thrust against your palm, as if you held a beating heart in a paper bag. What could I not do if I had the power and will of a mole! But the mole churns earth.

Last summer some muskrats had a den under this tree's roots on the bank; I think they are still there now. Muskrats' wet fur rounds the domed clay walls of the den and slicks them smooth as any igloo. They strew the floor with plant husks and seeds, rut in repeated bursts, and sleep humped and soaking, huddled in balls. These, too, are part of what Buber calls "the infinite ethos of the moment."

I am not here yet; I can't shake that day on the interstate. My mind branches and shoots like a tree.

Under my spine, the sycamore roots suck watery salts. Root tips thrust and squirm between particles of soil, probing minutely; from their roving, burgeoning tissues spring infinitesimal root hairs, transparent and hollow, which affix themselves to specks of grit and sip. These runnels run silent and deep; the whole earth trembles, rent and fissured, hurled and drained. I wonder what happens to root systems when trees die. Do those spread blind networks starve, starve in the midst of plenty, and desiccate, clawing at specks?

Under the world's conifers—under the creekside cedar behind where I sit—a mantle of fungus wraps the soil in a weft, shooting out blind thread after frail thread of palest dissolved white. From root tip to root tip, root hair to root hair, these filaments loop and wind; the thought of them always reminds me of Rimbaud's "I have stretched cords from steeple to steeple, garlands from window to window, chains of gold from star to star, and I dance." King David leaped and danced naked before the ark of the Lord in a barren desert. Here the very looped soil is an intricate throng of praise. Make connections; let rip; and dance where you can.

The insects and earthworms, moles, muskrats, roots and fungal strands are not all. An even frailer, dimmer movement, a pavane, is being performed deep under me now. The nymphs of cicadas are alive. You see their split skins, an inch long, brown, and translucent, curved and segmented like shrimp, stuck arching on the trunks of trees. And you see the adults occasionally, large and sturdy, with glittering black and green bodies, veined transparent wings folded over their backs, and artificial-looking, bright red eyes. But you never see the living nymphs. They are underground, clasping roots and sucking the sweet sap of trees.

In the South, the periodical cicada has a breeding cycle of thirteen years, instead of seventeen years as in the North. That a live creature spends thirteen consecutive years scrabbling around in the root systems of trees in the dark and damp—thirteen years!—is amply boggling for me. Four more years—or four less—wouldn't alter the picture a jot. In the dark of an April night the nymphs emerge, all at once, as many as eighty-four of them digging into the air from every square foot of ground. They inch up trees and bushes, shed their skins, and begin that hollow, shrill grind that lasts all summer. I guess as nymphs they never see the sun. Adults lay eggs in slits

along twig bark; the hatched nymphs drop to the ground and burrow, vanish from the face of the earth, biding their time, for thirteen years. How many are under me now, wishing what? What would I think about for thirteen years? They curl, crawl, clutch at roots and suck, suck blinded, suck trees, rain or shine, heat or frost, year after groping year.

And under the cicadas, deeper down than the longest taproot, between and beneath the rounded black rocks and slanting slabs of sandstone in the earth, ground water is creeping. Ground water seeps and slides, across and down, across and down, leaking from here to there minutely, at the rate of a mile a year. What a tug of waters goes on! There are flings and pulls in every direction at every moment. The world is a wild wrestle under the grass: earth shall be moved.

Camouflage and Disguise

Even when they are in the presence of a predator, many animals are able to go unnoticed—and thus unharmed—because their bodies are effectively masked by appropriate shades and arrangements of colors, patterns or shapes. Known as crypsis, this widespread means of defense is a form of concealment that is vital to creatures forced to be out in the open. Though crypsis takes a wide variety of forms, they fall into one of two broad categories: camouflage, in which colors and patterns either allow an animal to blend into its natural surroundings or break up the outline of its body into shapes unrecognizable to a predator; and disguise, in which an animal has evolved to resemble another specific object in its environment—a leaf, twig or bird dropping, for example—that is of no interest to a predator as food. In order to work, each technique must be accompanied by appropriate behavior: An animal must instinctively choose a background that will make it inconspicuous; it must have keen senses to detect a threat; and it must instantly fall silent, stop its normal activities, and either become motionless or imitate the natural movement of its setting, like leaves or reeds swaying in a breeze.

Camouflage is most prevalent among those creatures to which it is most essential. These are likely to be small, weak and otherwise defenseless animals: female birds and other animals that bear the burden of protecting and raising their young, and the eggs and young of animals that nest in exposed locations. The latter are likely to be more heavily stippled or mottled than the parents, giving them added protection at the most vulnerable stage.

One of the most intriguing aspects of camouflage is the way many species evolve different coloration to conform to the character of their environment. This phenomenon, called polymorphism, can be observed in many desert species. Populations of pocket mice living on black lava rock in New Mexico, for example, are nearly black, while in the nearby sandy desert, members of the same species are buff colored, and in the gypsum of White Sands National Monument they are almost pure white.

The colors of many creatures adapt to the changing seasons, increasing the animals' chances of staying alive. For example, northern mammals like the snowshoe rabbit, the ermine and the Arctic fox are brown in summer but in winter develop white coats that match the snow.

In contrast to these examples of animals that gradually change with the seasons, other animals have evolved as true quick-change artists. Various species of frogs, lizards and fish can change their colors in a matter of hours to blend with new surroundings. No other creature, however, can equal the octopus and its relatives, the cuttlefish and the squid. Octopuses can match any background ranging from black to white and various seaweed-and-rock combinations in between—in less than a second.

But, however versatile and adaptable, protective coloration is of little value if strong sunlight throws its owner's body outline into sharp relief or casts a revealing shadow on the ground. Many animals overcome this problem by "getting rid" of their shadows, or at least minimizing them—a phenomenon that has been called the "Peter Pan effect," a reference to the children's classic in which the hero loses his shadow. Moths press their wings close to tree trunks while resting during the day; antelope, deer, ground-dwelling birds and other creatures instinctively crouch or flatten against the ground when alarmed. Many, if not most, cryptic creatures also exhibit countershading—they are darker colored on their upper sides than on their lower ones to balance out the normal toplighting effect of the sun. Their bodies thus seem more uniformly colored and flat, and are more difficult to pick out against a background of similar color.

A more elaborate means of camouflage, often combined with countershading, is disruptive coloration, which means that a solid, recognizable body shape is broken up into distracting patches or stripes. The best-known example is the zebra, whose cryptic zigzag markings—wider and blacker above than below—tend to make it disappear, especially in the uncertain light of dawn and dusk when a lion is most likely to be out hunting.

Furthest advanced in the arts of concealment are those creatures that avoid attack by disguising themselves as something they are not. They have evolved into near-perfect replicas of bits of their natural environment: flowers, twigs, thorns, reeds or stones. Perhaps the most accomplished of all is the leaf fish of the Amazon Valley, whose disguise includes veinlike markings and even a stalklike growth protruding from its chin. It drifts just beneath the water's surface like a dead, waterlogged leaf, propelling itself occasionally with a ripple of its transparent fins. Larger fish overlook it, and so do smaller ones. When opportunity approaches, the leaf fish can come to life at the last moment and gobble up a meal.

Tree frog

34

Keeping a Low Profile

Camouflage is a defensive mechanism especially useful to those animals, such as nocturnal insects and spiders, that are inactive during the daylight hours when they would otherwise be highly visible to predators.

When a short-horned grasshopper is resting, predators are distracted and confused by the random color patches on its body. This pattern, called disruptive coloration, blurs the outline and texture of the grasshopper's body so that it becomes less recognizable to a predator. The leafhopper's measure of invisibility is heightened by striated body markings that look like the bark of a tree.

Camouflage is used by predators as well as by prey. The crab spider's color changes to match a variety of floral shades, a melding of flora and fauna called alluring coloration. The spider nestles down into a blossom, where it waits to ensnare insects drawn to the flower's nectar.

A leafhopper (above) crouches close to the surface of a tree limb to minimize the shadows that would attract predators.

Safely hidden from predators, a crab spider (left) uses a brightly colored blossom as a kind of improvised hunting blind.

Protected by cryptic coloring, a short-horned grasshopper (right) instinctively folds its wings close to its body to further obscure its presence.

Going Incognito

The art of disguise is a variation on the general camouflage theme, and many inanimate objects—flowers, leaves, twigs, thorns—serve as models for cryptic masqueraders, some of which are shown here and on the following pages.

The masters of cryptic disguise owe much of their success to their ability to freeze when threatened or when preparing to strike. A clumsy movement might reveal their presence to predators or prey—which, having then cracked the cryptic code, could establish a so-called searching image that would make it easier for them to detect their enemies. Members of similar cryptic species often separate themselves geographically from their fellows to cut down on the likelihood of having their code deciphered.

Among cryptic animals, flower mimics are particularly versatile; many change color to match the flower on which they have settled or, like a flower mantis, so resemble a blossom that other insects are lured to it.

Virtually nothing betrays the presence of the Malaysian flower mantis. The insect's front legs are carefully tucked up against its body.

A butterfly alights on a bright-colored flower bloom—only to be seized by the saw-toothed forelegs of a well-disguised African flower mantis.

A caterpillar cocoon (above) is veined
to simulate a flower pod—and shaped
to accommodate the wings of the
butterfly-to-be.

A chrysalis (above) is veined to
simulate a seedpod—and shaped to
accommodate the developing wings of
the butterfly-to-be.

Doing a handstand, a leaf katydid
(right) mimics the foliage surrounding
it—giving the impression of a partly
furled, slightly discolored leaf.

A spanworm caterpillar (left) does a
twig imitation, anchoring itself to a
branch with its hind appendages and
holding its head and thorax rigidly
away from the wood.

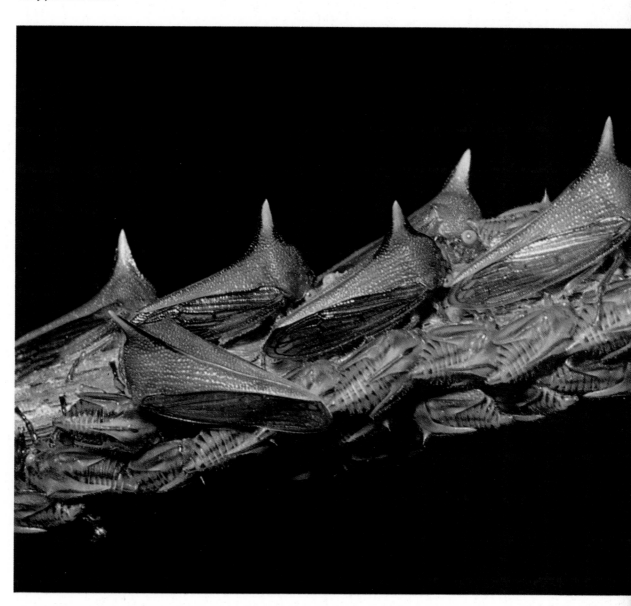

A treehopper nymph (above) disguised
as a green thorn, climbs up to join a
group of its prickly fellows at the
communal snack bar.

Diaphanous Delusions

Bright light is the nemesis of many cryptic species—it creates revealing shadows, betrays inadvertent movements and reveals the inevitable imperfections in apparently flawless cryptic stratagems. But for some fish and insect species bright light is a defensive advantage: Transparency is the solution. In shallow surface waters the light is unrelenting, and if a substantial part of a fish's body is crystal clear, light rays pass right through it. The light cannot clearly delineate the fish's body outline, or the textural qualities of its body surface. If such distinguishing features as internal organs are visible, they are often countershaded —colored so as to reflect light from above—for further protection. A predator looking toward a glass catfish, for example, will usually be rewarded only by a glimpse of the sun-dappled aquatic plant life behind the fish.

Many other marine species have also gone the transparency route, and these mollusks, crustaceans and various other invertebrates differ from their shore- or bottom-dwelling relatives in this important respect. Several members of the insect world have also evolved the see-through characteristic, which helps to divert attention from themselves to their surroundings.

Although the bone structure and internal organs of the scaleless glass catfish (below) are opaque, the fish's main body areas are limpid.

The clear green chrysalis of a glasswing butterfly (above, left) hangs suspended from a thread of gossamer that has surprising strength.

Like panes of leaded glass, the wings of a fulgaroid plant hopper (above, right) allow an almost unobstructed view of background foliage.

Piscine Costumery

Flounders and sand dabs, members of the flatfish family, are cryptic fish that have an evolutionary edge on many of their protectively colored marine counterparts in that they can also alter their body coloration to match changes in their environments.

These piscine chameleons have laterally flattened bodies that are well adapted to life close to the ocean floor in the relatively shallow areas of the continental shelf. They also have sensitive color vision that registers the subtlest color gradations in the sea bottom. This information is passed through their complex nervous system to chromatophores, skin cells containing pigment, which faithfully reproduce not only the colors of the ocean floor — ranging from pale sandy tones to deep brown—but also its texture. Within a few hours of arriving in a new location, the fish's body pattern, which may vary from speckled, mottled configurations to more uniform coloration, is in harmony with the mud, loose sand or stony seabed that conceals it.

When a flounder camouflages itself on the ocean floor, almost nothing can betray its presence except perhaps a flicker of its wary eyes.

Two speckled sand dabs drift down onto a gravelly sea floor. Some of their fins are equipped with fingerlike projections that hug the bottom.

44

A Rainbow Palette

The torpedo-shaped squids and their chunkier relatives, the cuttlefish, are the animal kingdom's quick-change camouflage champions—changing color in as little as two thirds of a second. Underneath their outer skin there are usually three layers of chromatophores, or pigment cells. The cells in each layer are composed of a different color— frequently red, yellow and brown. Triggered by impulses transmitted by the central nervous system, the muscle fibers attached to each chromatophore may contract, enlarging some of the elastic cell membranes and creating greater areas of pigmentation in one or more layers. Different sensory impulses may relax the muscle fibers, shrinking the chromatophores and blocking their colors.

These combinations of open and closed cells determine overall body coloration. The interaction among central nervous system, muscle fibers and pigment cells can produce primary colors or combine them to create secondary ones. The result can be a dramatic light show or a more subdued mottled effect. The complex mechanism allows squid and cuttlefish to match their colorations to their backgrounds with spectacular accuracy. But its workings remain a mystery; marine biologists have yet to figure out how the mollusks' primitive central nervous systems interpret visual perceptions of color and then translate these interpretations into the specific impulses that trigger appropriate color changes.

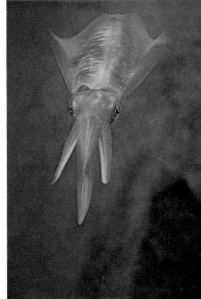

Like a bejewelled pelagic phantom, an opalescent squid (right, above) gleams with a milky-white radiance: Its red, yellow and brown pigment cells are contracted, exposing a white area that consists of iridescent reflecting cells.

A harmless-looking cuttlefish (right) is a well-camouflaged death trap for tiny crustaceans: Its sharp and powerful beak is obscured by the eight waving tentacles that surround it. Two hidden retractable tentacles dart out to seize the prey, then retract again.

Glowing like an incandescent rocket, a cuttlefish (left) arranges its tentacles to achieve a perfect three-point landing on the ocean bottom. The luminescent color of the cryptic mollusk blends artfully with the seabed on which it is about to settle.

Rippling waves of color play over the body of a lurking cuttlefish (right). Fascination with these impressive visual effects has been the downfall of many a tiny crustacean: Tentacles, tipped with suckers, suddenly flick out to grab the distracted victim.

Changes of Hue

Sea hares are marine mollusks related to the squids and cuttlefish, sharing with them their defensive abilities to emit an opaque fluid and to change their body coloration. Though equally mysterious, the sea hare's cryptic technique is more orderly than that of its flashier mollusk cousins and follows a predictable course closely attuned to the life cycle of the species.

As larvae, these gastropods eat the rose-red algae that flourish beyond the low-tide level, and at this stage their body coloration matches that deep red. As they mature, the young sea hares migrate shoreward. Their coloration gradually changes to match the rich olive green of the littoral algae that become their new food supply. Experiments have proved that sea hares do not absorb the pigmentation of the algae they feed on, and the causes of their radical color change remain unknown.

Entrenched in a cranny, a sea hare bears a striking resemblance to a cluster of marine vegetation—in a color scheme that duplicates that of the plants it eats.

A sea hare (right) ejects a milky secretion that confuses and often discourages a predator. The emission of this harmless purple fluid is merely an intimidating action.

Prey and Predator

For many animals, such as the highly vulnerable tree frogs of Central and South America, crypsis is a primary line of defense. But it can also be an efficacious offensive tool, used by some of nature's most ferocious predators—including the alligator shown on the following pages. The alligator, a reptile, is at the opposite end of the cryptic spectrum: a well-camouflaged hunter with few natural enemies—other than man—rather than a protectively colored potential victim.

Their protective coloration adapts tree frogs to particular environments. Different tree frogs may be gray, brown, yellow or green. All can become darker or lighter, and some can actually change their color. Hormonal agents carried in the bloodstream produce these color changes within a matter of a few hours. The red-eyed tree frog is small enough to hide within a slightly furled leaf. Outside its native jungle habitat its vivid coloration would be most conspicuous, but in the lush green rain forest, the diminutive amphibian is easily overlooked by the snakes and birds that prey upon the species.

Unlike fleet-footed animals that count on speed to run down and overcome their prey and do not depend heavily on body camouflage, the alligator, a lethargic creature, must lie in wait to ensnare passing victims. The big reptile's cryptic coloration is vital to the success of this passive hunting technique.

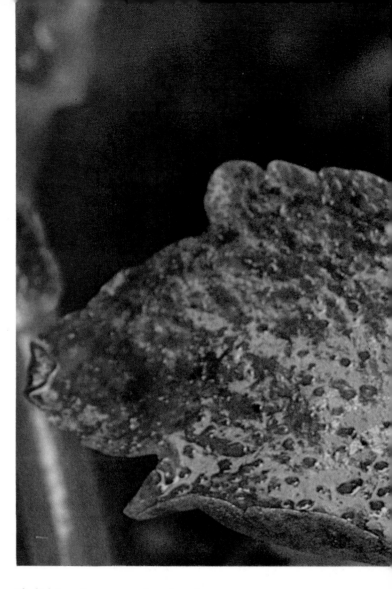

The bulging crimson eyes of a red-eyed tree frog (below) somewhat offset its otherwise highly successful camouflage technique.

The red-eyed tree frog's effort to conceal itself succeeds only after its flashing orbs are discreetly veiled by thin, translucent eyelids (below).

A glistening Venezuelan tree frog (above) flattens itself on the dewy surface of a leaf, which provides cryptic concealment.

An alligator (overleaf) drifts slowly, looking like a scabrous chunk of dead wood. Only its cold, glassy eyes betray the presence of the cruising killer.

Avian Artifices

Protective coloration is an essential survival mechanism for both arboreal birds and those that live and nest on the ground; the latter are especially vulnerable. The ptarmigan, a terrestrial bird, is subject to all the hazards found on the ground. Molting enables the ptarmigan to change colors, matching seasonal variations in the ground cover. The subdued fall plumage gradually whitens as the cold weather intensifies, and eventually rivals the winter snow in brilliance. When the snow melts in spring, the bird molts

again, developing a patchy plumage that blends with the partially exposed ground. The male often molts later than the female; his white feathers catch the eye of predators, distracting their attention from his incubating mate.

The osprey—a raptor—usually builds its large nest high above the ground in trees, on rocky promontories or even on telephone poles. Although adult ospreys have no natural enemies except man, their nestlings are vulnerable and therefore are protectively patterned.

Four white-tailed ptarmigans (left) brave the cold of a winter wasteland. Instinctively huddled together near a clump of bare branches, the birds look like nothing more than random drifts of windblown snow.

Newly hatched, cryptically patterned rock ptarmigans cluster together amid a jumble of broken eggshells. The smooth white inner surfaces of the shells might attract the attention of predators, so they are quickly removed from the nest by the hen.

At a warning note from their mother, two osprey fledglings freeze in position. With wings partially extended and heads drooping over the side of the nest, the young ospreys will feign death until another maternal signal indicates that danger is past.

55

How the Leopard Got His Spots

by Rudyard Kipling

Among the many works of Rudyard Kipling that were inspired by his experiences in India, the most enduring have been his books for children—Kim, The Jungle Books, *and* Just So Stories, *first published in 1902, which contributes the tale reprinted here with Kipling's own illustration. Although "How the Leopard Got His Spots" recognizes the importance of camouflage to wild animals, Kipling's delightful explanation of the defense's origins was, of course, meant only to entertain.*

In the days when everybody started fair, Best Beloved, the Leopard lived in a place called the High Veldt. 'Member it wasn't the Low Veldt, or the Bush Veldt, or the Sour Veldt, but the 'sclusively bare, hot, shiny High Veldt, where there was sand and sandy-coloured rock and 'sclusively tufts of sandy-yellowish grass. The Giraffe and the Zebra and the Eland and the Koodoo and the Hartebeest lived there; and they were 'sclusively sandy-yellow-brownish all over; but the Leopard, he was the 'sclusivest sandiest-yellowest-brownest of them all—a greyish-yellowish catty-shaped kind of beast, and he matched the 'sclusively yellowish-greyish-brownish colour of the High Veldt to one hair. This was very bad for the Giraffe and the Zebra and the rest of them; for he would lie down by a 'sclusively yellowish-greyish-brownish stone or clump of grass, and when the Giraffe or the Zebra or the Eland or the Koodoo or the Bush-Buck or the Bonte-Buck came by he would surprise them out of their jumpsome lives. He would indeed! And, also, there was an Ethiopian with bows and arrows (a 'sclusively greyish-brownish-yellowish man he was then), who lived on the High Veldt with the Leopard; and the two used to hunt together—the Ethiopian with his bows and arrows, and the Leopard 'sclusively with his teeth and claws—till the Giraffe and the Eland and the Koodoo and the Quagga and all the rest of them didn't know which way to jump, Best Beloved. They didn't indeed!

After a long time—things lived for ever so long in those days—they learned to avoid anything that looked like a Leopard or an Ethiopian; and bit by bit—the Giraffe began it, because his legs were the longest—they went away from the High Veldt. They scuttled for days and days and days till they came to a great forest, 'sclusively full of trees and bushes and stripy, speckly, patchy-blatchy shadows, and there they hid: and after another long time, what with standing half in the shade and half out of it, and

what with the slippery-slidy shadows of the trees falling on them, the Giraffe grew blotchy, and the Zebra grew stripy, and the Eland and the Koodoo grew darker, with little wavy grey lines on their backs like bark on a tree trunk; and so, though you could hear them and smell them, you could very seldom see them, and then only when you knew precisely where to look. They had a beautiful time in the 'sclusively speckly-spickly shadows of the forest, while the Leopard and the Ethiopian ran about over the 'sclusively greyish-yellowish-reddish High Veldt outside, wondering where all their breakfasts and their dinners and their teas had gone. At last they were so hungry that they ate rats and beetles and rock-rabbits, the Leopard and the Ethiopian, and then they had the Big Tummy-ache, both together; and then they met Baviaan—the dog-headed, barking Baboon, who is Quite the Wisest Animal in All South Africa.

Said Leopard to Baviaan (and it was a very hot day), 'Where has all the game gone?'

And Baviaan winked. *He* knew.

Said the Ethiopian to Baviaan, 'Can you tell me the present habitat of the aboriginal Fauna?' (That meant just the same thing, but the Ethiopian always used long words. He was a grown-up.)

And Baviaan winked. *He* knew.

Then said Baviaan, 'The game has gone into other spots; and my advice to you, Leopard, is to go into other spots as soon as you can.'

And the Ethiopian said, 'That is all very fine, but I wish to know whither the aboriginal Fauna has migrated.'

Then said Baviaan, 'The aboriginal Fauna has joined the aboriginal Flora because it was high time for a change; and my advice to you, Ethiopian, is to change as soon as you can.'

That puzzled the Leopard and the Ethiopian, but they set off to look for the aboriginal Flora, and presently, after ever so many days, they saw a great, high, tall forest full of tree trunks all 'sclusively speckled and sprottled and spottled, dotted and splashed and slashed and hatched and cross-hatched with shadows. (Say that quickly aloud, and you will see how *very* shadowy the forest must have been.)

'What is this,' said the Leopard, 'that is so 'sclusively dark, and yet so full of little pieces of light?'

'I don't know,' said the Ethiopian, 'but it ought to be the aboriginal Flora. I can smell Giraffe, and I can hear Giraffe, but I can't see Giraffe.'

'That's curious,' said the Leopard. 'I suppose it is because we have just come in out of the sunshine. I can smell Zebra, and I can hear Zebra, but I can't see Zebra.'

'Wait a bit,' said the Ethiopian. 'It's a long time since we've hunted 'em. Perhaps we've forgotten what they were like.'

'Fiddle!' said the Leopard. 'I remember them perfectly on the High Veldt, especially their marrow-bones. Giraffe is about seventeen feet high, of a 'sclusively fulvous golden-yellow from head to heel; and Zebra is about four and a half feet high, of a 'sclusively grey-fawn colour from head to heel.'

'Umm,' said the Ethiopian, looking into the speckly-spickly shadows of the aboriginal Flora-forest. 'Then they ought to show up in this dark place like ripe bananas in a smoke-house.'

But they didn't. The Leopard and the Ethiopian hunted all day; and though they could smell them and hear them, they never saw one of them.

'For goodness' sake,' said the Leopard at tea-time, 'let us wait till it gets dark. This daylight hunting is a perfect scandal.'

So they waited till dark, and then the Leopard heard something breathing sniffily in the starlight that fell all stripy through the branches, and he jumped at the noise, and it smelt like Zebra, and it felt like Zebra, and when he knocked it down it kicked like Zebra, but he couldn't see it. So he said, 'Be quiet, O you person without any form. I am going to sit on your head till morning, because there is something about you that I don't understand.'

Presently he heard a grunt and a crash and a scramble, and the Ethiopian called out, 'I've caught a thing that I can't see. It smells like Giraffe, and it kicks like Giraffe, but it hasn't any form.'

'Don't you trust it,' said the Leopard. 'Sit on its head till the morning—same as me. They haven't any form—any of 'em.'"

So they sat down on them hard till bright morning-time, and then Leopard said, 'What have you at your end of the table, Brother?'

The Ethiopian scratched his head and said, 'It ought to be 'sclusively a rich fulvous orange-tawny from head to heel, and it ought to be Giraffe; but it is covered all over with chestnut blotches. What have you at *your* end of the table, Brother?'

And the Leopard scratched his head and said, 'It ought to be 'sclusively a delicate greyish-fawn, and it ought to be Zebra; but it is covered all over with black and purple stripes. What in the world have you been doing to yourself, Zebra? Don't you know that if you were on the High Veldt I could see you ten miles off? You haven't any form.'

'Yes,' said the Zebra, 'but this isn't the High Veldt. Can't you see?'

'I can now,' said the Leopard. 'But I couldn't all yesterday. How is it done?'

'Let us up,' said the Zebra, 'and we will show you.'

They let the Zebra and the Giraffe get up; and Zebra moved away to some little thornbushes where the sunlight fell all stripy, and Giraffe moved off to some tallish trees where the shadows fell all blotchy.

'Now watch,' said the Zebra and the Giraffe. 'This is the way it's done. One—two—three! And where's your breakfast?'

Leopard stared, and Ethiopian stared, but all they could see were stripy shadows and blotched shadows in the forest, but never a sign of Zebra and Giraffe. They had just walked off and hidden themselves in the shadowy forest.

'Hi! Hi!' said the Ethiopian. 'That's a trick worth learning. Take a lesson by it, Leopard. You show up in this dark place like a bar of soap in a coal-scuttle.'

'Ho! Ho!' said the Leopard. 'Would it surprise you very much to know that you show up in this dark place like a mustard-plaster on a sack of coals?'

'Well, calling names won't catch dinner,' said the Ethiopian. 'The long and the little of it is that we don't match our backgrounds. I'm going to take Baviaan's advice. He told me I ought to change; and as I've nothing to change except my skin I'm going to change that.'

'What to?' said the Leopard, tremendously excited.

'To a nice working blackish-brownish colour, with a little purple in it, and touches of slaty-blue. It will be the very thing for hiding in hollows and behind trees.'

So he changed his skin then and there, and the Leopard was more excited than ever; he had never seen a man change his skin before.

'But what about me?' he said, when the Ethiopian had worked his last little finger into his fine new black skin.

'You take Baviaan's advice too. He told you to go into spots.'

'So I did,' said the Leopard. 'I went into other spots as fast as I could. I went into this spot with you, and a lot of good it has done me.'

'Oh,' said the Ethiopian, 'Baviaan didn't mean spots in South Africa. He meant spots on your skin.'

'What's the use of that?', said the Leopard.

'Think of Giraffe,' said the Ethiopian. 'Or if you prefer stripes, think of Zebra. They find their spots and stripes give them per-fect satisfaction.'

'Umm,' said the Leopard. 'I wouldn't look like Zebra—not for ever so.'

'Well, make up your mind,' said the Ethiopian, 'because I'd hate to go hunting without you, but I must if you insist on looking like a sun-flower against a tarred fence.'

'I'll take spots, then,' said the Leopard; 'but don't make 'em too vulgar-big. I wouldn't look like Giraffe—not for ever so.'

'I'll make 'em with the tips of my fingers,' said the Ethiopian. 'There's plenty of black left on my skin still. Stand over!'

Then the Ethiopian put his five fingers close together (there was plenty of black left on his new skin still) and pressed them all over the Leopard, and wherever the five fingers touched they left five little black marks, all close together. You can see them on any Leopard's skin you like, Best Beloved. Sometimes the fingers slipped and the marks got a little blurred; but if you look closely at any Leopard now you will see that there are always five spots—off five fat black finger-tips.

'Now you *are* a beauty!' said the Ethiopian. 'You can lie out on the bare ground and look like a heap of pebbles. You can lie out on the naked rocks and look like a piece of pudding-stone. You can lie out on a leafy branch and look like sunshine sifting through the leaves; and you can lie right across the centre of a path and look like nothing in particular. Think of that and purr!'

Woodland Crypsis

Born with a spotted coat, the young white-tailed deer is almost indistinguishable from the dense, sun-dappled vegetation on which it lies. But the adult whitetail loses this advantage along with its spots, and depends for survival on elusive behavior. The whitetail is usually a loner that seldom ventures very far from the woods or from other areas of dense growth that provide it protection. When danger threatens, it crouches and maintains an immobile position as close to the earth as possible, its head and neck held parallel to the ground. By assuming this posture, the deer minimizes shadows that might reveal it and reduces its visible body area.

Rapid flight is the last line of defense for the adult white-tailed deer. The young, however, must rely on concealment. When the doe has established her offspring in an out-of-the-way clump of tall grass or undergrowth, she is free to concentrate on foraging for food. But until her youngster is about six weeks old, she returns every four hours or so to nurse it.

A white-tailed buck (above), flushed from its woodland retreat, stands at bay on a snowy slope. A more cautious fellow whitetail (right) remains partly protected by a screen of tall grass.

Warning Signals

Most people who have been stung by a bee or a wasp keep a wary eye on flying objects that buzz and have black-and-yellow stripes. What is more, they usually steer clear of similar-looking insects, even though it is not certain that they will sting. Like other animals, humans have learned by experience to avoid a creature that has a dangerous or unpleasant attribute as its main line of defense. Among the many animals that have stings, poisonous secretions, foul odors or evil-tasting flesh, most are able to keep these mechanisms in reserve—and stay out of trouble—advertising their disagreeable qualities by means of colors and other signals that warn potential attackers to beware. This form of defense, called aposematism, is the exact opposite of hiding and camouflage, and in its way it works just as well. Almost without exception, animals equipped with noxious weapons—hornets, stinkbugs and bombardier beetles; skunks and porcupines; jellyfish and sea anemones; poisonous lizards, fishes and frogs—are brilliantly colored and patterned so that they stand out from their surroundings instead of blending in. Like stop signs, these patterns are readily recognizable by any animal that has once sampled the owner's wares. They read "Danger," "Stay Away," "Don't Fool Around with Me."

To make sure the message is understood, aposematic animals act in distinctly different ways from cryptic and anachoretic ones. Their movements are actually designed to make them conspicuous. Instead of darting away when threatened, they go about their activities deliberately, as if daring other animals to catch them. The purpose of this behavior is instructional: Since few animals instinctively shy away on first meeting from a species that uses color as a warning, some individuals must expose themselves to attack so that uninitiated predators may learn about them and subsequently avoid their kin.

Unlike many cryptic species, most aposematic animals are active by day when their warning colors are most easily perceived; the few that are nocturnal, like badgers and skunks, have no need for bright hues—most nocturnal predators are color-blind anyway—but instead tend to rely on sharply contrasting black-and-white markings that advertise such traits as foul odor and ferocity, and can be seen clearly on all but pitch-black nights.

Some aposematic species living in the same region resemble one another, often in minute detail as far as their highly visible colors, markings and shapes are concerned.

This is known as Müllerian mimicry, and was first discovered by a German naturalist, Fritz Müller, who noticed that different species of Brazilian butterflies, all unpalatable to birds and lizards, were almost indistinguishable from one another until examined in detail. The value of membership in such a "warning club" soon became apparent: Instead of sampling many different individuals, predators had only to learn one general combination of pattern and color by experience in order for all the club members to benefit. Müllerian mimicry, an intriguing by-product of natural selection, is particularly widespread among insects. A common example is that of the general black-and-yellow coloration of many bees and wasps. And birds that have learned to avoid one type of wasp will also avoid similar representatives of other wasp species.

An even more interesting twist in animal resemblances is the case of non-noxious species that look almost exactly like noxious ones, and thus gain membership in the warning club without having to pay any dues. Called Batesian mimicry, this phenomenon was first discovered by an English naturalist, H. W. Bates, while studying butterflies in Brazil. Batesian mimics are, in effect, sheep in wolves' clothing. Various flies, moths and beetles have evolved into convincing Batesian mimics of hornets, bees and wasps; though harmless, they not only display black-and-yellow markings but also imitate the flight style of their models and buzz angrily, particularly when alarmed. Some small spiders mimic ants, apparently to escape attack by other spiders. Mingling with their adopted protectors, which defend themselves by secreting formic acid, the "guest" spiders scurry about on six of their legs while holding the other two, like an ant's antennae, in the air.

A fascinating example of the lengths to which mimicry can go is that of some 70 species of New World coral snakes, all of which have bodies banded in various combinations and shadings of red, yellow, black and white. Some are harmless Batesian mimics of venomous snakes, while others are poisonous Müllerian mimics of one another. Still others are so-called Mertensian mimics—named after R. Mertens, a scientist who studied New World coral snakes extensively. Highly venomous species, these coral snakes enhance their protection by actually imitating less poisonous snakes, using the mildly toxic species as demonstrators to educate predators, which their own strong venom would not teach but would kill.

Solomon Islands spiny-backed spider

Code for "Nasty"

Although aposematic animals come in myriad shapes and sizes, virtually all of them share one basic characteristic: a bold color scheme designed to advertise a disagreeable or dangerous trait—such as noxious or foul-tasting secretions, poisonous stings or spines—that makes the animal unpalatable or even inedible. When an inexperienced predator fails to heed such an animal's warning colors, a whiff or a nibble are often all that is needed to spring the prey—and make the predator more careful the next time it encounters a member of that species.

Most warning patterns consist of some combination of red, orange, yellow, black and white, which are highly visible against the generally green and brown hues found in nature. The most effective of these color schemes are the simplest ones—a few colors arranged in a striking pattern. The less ornate the design, the easier it is for a predator to recognize and remember it as the signature of an unpalatable species.

The dazzling pink of the nymph of the Bornean stinkbug Pycanum rubeus *(above) alerts passersby to its ability to produce a noxious secretion.*

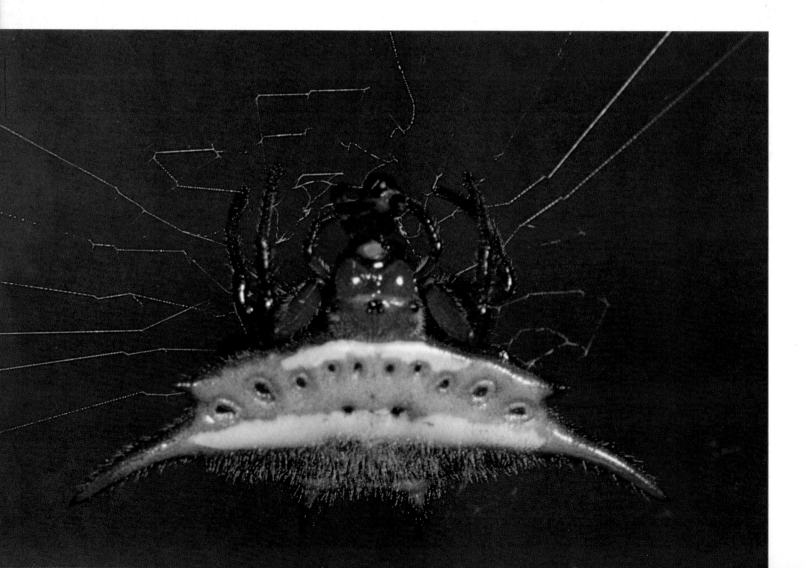

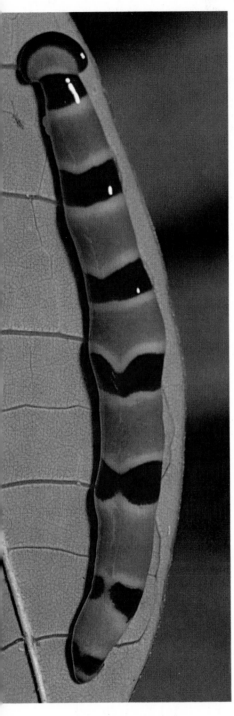

The stripes of this Malaysian
turbellarian worm (above) are
impressive to any predator that has
sampled its nauseating flesh.

East Africa is home to the fire-engine-
red and yellow spider Gasteracantha
falcicornis (left). The alarming colors of
this tree dweller warn of a formidable
array of abdominal spines.

An aposematic grasshopper, Zonoceros
elegans, sits fearlessly on a leaf. If its
bright colors fail to fend off enemies, it
can drive them away by discharging a
noxious foam.

Wild Season
by Allan M. Eckert

Allan M. Eckert, an experienced wilderness traveler and observer since childhood, has written many articles and children's books dealing with natural history. In this passage from Wild Season, *published in 1967, the month of May brings a frog out of hibernation to search for food. A black ground beetle appears palatable enough to the frog, but like many insects, it has a defense that guarantees that no predator will mistake it for a tempting morsel a second time.*

Already he had come thirty yards or more from where he had emerged yesterday from hibernation, and the ground here was much more solid, covered with a new growth of land plants rather than aquatics or marsh growth. There were trees here, too, the first he had ever been close to, and the base of each of these he found to be a good hunting ground for smaller insects. Ants, spiders, flies, bees, caterpillars and other invertebrates seemed drawn to them, and as he crouched beneath one tree a little black ground beetle lumbered nonchalantly toward him. He watched its approach attentively and when it was within catching distance he leaped forward and snapped it in.

At once a nauseous material filled his mouth and the accompanying stench was sickening. Hastily he spat out the offensive insect, which, only a little the worse for wear, wobbled away into the undergrowth. The frog clawed at his mouth and tongue with his front feet and his eyes

blinked rapidly. After a moment he regurgitated the residue of insects from his stomach, but still the terrible taste and burning remained.

At length, his eyes smarting and skin badly irritated by the fluid the ground beetle had ejected, the little bullfrog leaped to the edge of the bank, poised there a moment and then dived into the water. He kicked his way to the sandy bottom, which was deeper here than along the marshy shoreline, and rubbed his body in the bottom grit, then even scooped up a mouthful and spat it out. He scraped along the bottom all the way back to shore and then headed for a little clump of blossoming water grass growing out of inch-deep water. Another lesson had been learned: certain smaller creatures than he possessed powerful weapons, and he would not again strike at little black beetles lest they be armed with such a noxious defense.

Dangerous Beauty

Among the most familiar and most beautiful of the 100,000 or more known species in the order Lepidoptera are the tiger moths, well known for the spectacular beauty of their warning coloration. They are named for the orange-and-black or yellow-and-black patterns of stripes, spots or patches displayed by many of the tiger moth species, which include the familiar garden tiger moth, a handsome creature common to Europe and North America.

Such color combinations in bold patterns appear frequently in the animal world to warn predators of unpleasant tastes or smells. This is certainly the function of the striking coloration of the garden tiger moth, which, like many other members of the family Arctiidae, is pretty but poisonous. Certain glands produce powerful toxins that flow through the insect's bloodstream. Other glands located in its thorax secrete poisonous bubbles that carry an unpleasant warning smell. The garden tiger moth thus exemplifies a principle learned by many predators: In general, if a creature is both brightly colored and highly visible, the odds are that it is unfit to eat.

Even when resting, the garden tiger moth (above) is an attention getter. But when it spreads its fore wings in a defensive display (right), its stunning orange-and-black hind wings make the moth truly resplendent.

Hairs and Spines

New England folk wisdom has it that the ratio of black to brown hairs in autumn on the hairy caterpillars called woolly bears can be used to predict the severity of the following winter. But there are many species of hair-covered caterpillars around the world, including those in the gallery on these pages, whose high degree of visibility warns not of bad weather but of the bad tastes, smells and stings so many of them can produce. While not all brightly colored hairy caterpillars are dangerous, many of them do have poison glands at the base of their spiny hairs. Caterpillars of the South American moths belonging to the genus *Automeris* are especially noxious, and their stings can cause temporary paralysis in human beings. So painful is the sting that the flannel moth caterpillars of Brazil have been given the name *bizos de fuero*, or "fire beasts."

Also disagreeable are the caterpillars of the lasiocampid moths, whose spiny hairs are arranged in small colorful patches hidden beneath folds of skin on their backs. When this caterpillar is disturbed, it arches its back in a porcupine-like attempt to drive these sharp, pointed hairs into its predator.

Many species of hairy caterpillars tend to congregate with their own kind in colorful colonies, making their warning message even more obvious than that of an individual. Some are capable of shedding sections of their irritating hairs, which fill the air around them and further discourage intruders.

Lasiocampid caterpillar

Caterpillars of the Malay lacewing butterfly

Saturniid moth caterpillar

Rose slug caterpillar

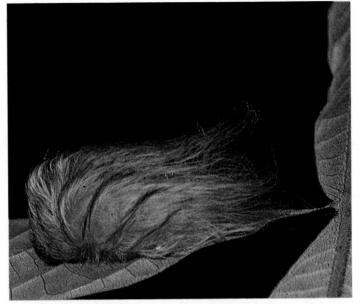

Flannel moth caterpillar

Tiger moth caterpillar

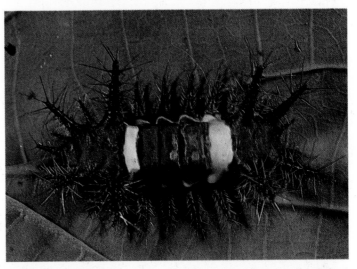

Eucleid moth caterpillar   Automeris io moth caterpillar (below)

Amphibious Showoffs

Although the vast majority of frogs and toads are nocturnal and depend primarily on cryptic coloration for defense—employing nature's shades of brown or green—there are several amphibian groups that include species that behave very differently. They are flamboyantly colored, sometimes sluggish in their movements, lead diurnal lives and make no attempt to hide from predators, unlike their more numerous camouflaged kin, which generally forage for food by night when they are best able to avoid predators.

Among the most conspicuous of the showoffs are the poison-dart frogs of the dendrobatid family of Central and South America, which secrete a powerful, paralyzing poison in their skin glands. A predator that has tasted a poison-dart frog and survived henceforth associates the unpleasant experience with these frogs' flashy colors and leaves the amphibians strictly alone.

A turquoise-and-orange poison-dart frog, Dendrobates granuliferus (above), stands out dramatically against the muted tones of a dried leaf.

The forests of northern South America are the home of the handsome poison-dart frog Dendrobates leucomelas (left).

Like most of its relatives, this red-and-black poison-dart frog, Dendrobates lehmanni (right), is quite small in size. Its habitat is terrestrial.

Shoulder to shoulder, a trio of young
skunks parade through a field
of dandelions. Young skunks are as
intrepid as their elders, holding their
ground rather than running and hiding
in the face of danger.

Striped Bravado

Most aposematic animals are diurnal, and their vivid warning coloration obviously has its strongest effect during daylight hours. Some species, however, such as most North and South American skunks, become active after dark; although their diet during the summer months includes some vegetable matter, the animals on which they feed —primarily insects, worms, rodents and other small prey—are more readily available at night. But the skunk's bold black-and-white pattern is nonetheless an effective defense mechanism. Even under cover of darkness its white stripes are visible.

To give an enemy fair warning, the skunk first stamps its forefeet and then raises its tail like a signal flag, exposing the anal glands from which it may finally eject a putrid-smelling fluid. The skunk appears to be aware of its considerable powers, for it goes about its business in a leisurely manner seemingly unafraid of other animals. This air of indifference belies the skunk's potentially savage disposition: If pushed into a confrontation, it can respond with both force and tenacity.

A skunk and a coyote (right) share a snowy stretch of prairie in Wyoming. The skunk protects itself by turning its back on a predator. In this position it can accurately aim its spray at an object up to 10 feet away.

Jim, the Story of a Backwoods Police Dog

by Charles G. D. Roberts

Born in Canada in 1860, Charles G. D. Roberts was among the earliest writers to derive literary themes from that country's vast landscapes. A poet and novelist as well as the author of many children's books portraying life in the wild, he is known as the father of Canadian letters. In the following selection from Jim, the Story of a Backwoods Police Dog, *the hero, a city canine, learns that a skunk's bold black-and-white pattern advertises a potent defense.*

Knowing his utter inability to compete with the speed of the rabbits, now they were wide awake, the skunk hardly noticed their antics, but kept on his direct path toward the farmyard. Presently, however, his attention was caught by the rabbits scattering off in every direction. On the instant he was all alert for the cause. Mounting a hillock, he caught sight of a biggish shaggy-haired dog some distance down the pasture. The dog was racing this way and that as crazily, it seemed, as the rabbits, with faint little yelps of excitement and whines of disappointment. He was chasing the rabbits with all his energy; and it was evident that he was a stranger, a new-comer to the wilderness world, for he seemed to think he might hope to catch

the fleet-foot creatures by merely running after them. As a matter of fact, he had just arrived that same day at the backwoods farm from the city down the river. His experience had been confined to streets and gardens and the chasing of cats, and he was daft with delight over the spacious freedom of the clearings. The skunk eyed him scornfully, and continued his journey with the unconcern of an elephant.

A moment later the dog was aware of a little, insignificant black-and-white creature coming slowly towards him as if unconscious of his presence. Another rabbit! But as this one did not seem alarmed, he stopped and eyed it with surprise, his head cocked to one side in inquiry. The skunk half turned and moved off slowly, deliberately, at right angles to the path he had been following.

With a yelp of delight the dog dashed at this easy victim, which seemed so stupid that it made no effort to escape. He was almost upon it. Another leap and he would have had it in his jaws. But the amazing little animal turned its back on him, stuck its tail straight in the air, and jerked up its hindquarters with a derisive gesture. In that instant something hot and soft struck the inexperienced hunter full in the face—something soft, indeed, but overwhelming, paralyzing. It stopped him dead in his tracks. Suffocating, intolerably pungent, it both blinded him and choked him. His lungs refused to work, shutting up spasmodically. Gasping and gagging, he grovelled on his belly and strove frantically to paw his mouth and nostrils clear of the dense, viscous fluid which was clogging them. Failing in this, he fell to rooting violently in the short grass, biting and tearing at it and rolling in it, till some measure of breath and eyesight returned to him.

Thereupon, his matted head all stuck with grass and moss and dirt, he set off racing madly for the farmhouse where he expected to get relief from the strange torment which afflicted him. But when he pawed and whined at the kitchen door for admittance, he was driven off with con-

tumely and broomsticks. There was nothing for him to do but slink away with his shame to a secluded corner between the wagon-shed and the pig-pen, where he could soothe his burning muzzle in the cool winds and fresh earth. On the following day one of the farm hands, with rude hands and unsympathetic comment, scrubbed him violently with liquid soap and then clipped close his splendid shaggy coat. But it was a week before he was readmitted to the comfortable fellowship of the farmhouse kitchen.

Visual Tricks

The concept of animal mimicry was first formulated in 1861 by the British naturalist H. W. Bates when he made a curious observation about the Brazilian butterflies he had been studying: Many of the bright-colored insects, though they belonged to different families, had similar color and behavior patterns. This led Bates to theorize, rightly, that at least one species of butterfly was distasteful to predators, while the others, though all too palatable, so closely resembled the undesirable model in color and markings that they were also left alone. Since Bates made this observation scores of other animals—some of which are shown here and on the following pages—have been found to benefit from the adaptation, which is known as Batesian mimicry.

One of the few known Batesian mimics of the fish world is the Indo-Pacific reef-dwelling plesiopid fish. When threatened, this otherwise defenseless creature darts headfirst into a crevice in rock or coral, exposing nothing but its rear end, which has a conspicuous ocellus, or eyespot. In this position the hind parts of the plesiopid look convincingly like the head of the dangerous moray eel. In the reptile world the poisonous coral snakes have numerous mimics—some dangerous, some not. All, however, are protected by variations of the coral snake's red, yellow and black bands.

The deadly white-spotted, brown-bodied moray eel Gymnothorax meleagris (left, top) of Hawaiian waters assumes its characteristic daytime posture, poking its head out from a crevice. At the first sign of danger, the similarly patterned plesiopid fish (left, bottom) goes headfirst into a niche. When its rear-facing anal, caudal and dorsal fins are expanded to reveal its eyespot, the fish strongly resembles the eel, and the effect is convincing enough to scare most predators away.

The order of the colors on the bodies of reptiles like the Eastern coral snake (right, top) inspired a rule of thumb for distinguishing venomous from nonvenomous species in the United States: "If red touch yellow, kill a fellow." At first glance the boldly banded Arizona king snake (right, bottom) would also be considered dangerous, a fact that contributes to its safety. But a closer look reveals black bands between the red and yellow. The verdict: nonpoisonous.

The handsome but foul-tasting monarch butterfly (above) owes its unpalatability to the fact that as a caterpillar it feeds on species of milkweed plants known to be poisonous to cattle and other vertebrates. Some of these plants contain cardenolides, substances that cause severe vomiting. Avian predators soon learn to avoid not only the monarch but also its only slightly distasteful mimic, the viceroy (below).

Striped for warning purposes but harmless, a hover fly (at left in the photograph above) extracts nectar from a flower, as do its two stinging honeybee models. The hover fly's mimicry involves not only its appearance but also its behavior. Hover flies imitate the sounds that bees and wasps make, buzzing in a threatening manner when disturbed. The entire act results in almost foolproof protection for the hover fly.

81

Stolen Weaponry

While many innocuous animals have evolved a re-semblance to noxious species and thus enjoy the same pro-tection afforded their models, there is another group of animals that assimilate the dangerous attributes of the crea-tures on which they prey. One such animal is the nudi-branch *Glaucus atlanticus*, a marine mollusk. *Glaucus* feeds on the tentacles of the Portuguese man-of-war, which contain stinging cells called nematocysts. Probably by means of a chemical reaction that is not yet understood, *Glaucus* is able to immobilize these cells, then eat and di-gest them intact. The nematocysts are incorporated into the finger-like projections on the mollusk's body where the sting is reactivated. The mollusk is then armed with a weapon—originally belonging to its coelenterate prey—that it can in turn use in its own defense.

A Portuguese man-of-war (above), its stinging tentacles drifting, floats on the surface. The tentacles protect the mollusk Glaucus (right) as it feeds on them. Once eaten, the stinging cells become part of Glaucus' defense system.

Bluffs and Threats

When an animal's first line of defense has been penetrated—whether its specialty is hiding, camouflage, warning colors or flight—it often falls back on another stratagem that intimidates or startles its attacker into confusion or retreat. Such action, called deimatic behavior after the Greek word for "frightening," includes a rich variety of techniques. Some are warning threats of genuine injury; others are purest bluff.

Probably the most familiar tactic is a conspicuous displaying of weapons: baring teeth, pawing the ground with sharp hooves or lowering pointed horns. Advertising their ammunition, porcupines erect their quills, skunks raise their tails in readiness to spray, and scorpions arch their stingers forward over their backs. Some animals also attempt to create an illusion of greater size by raising a ruff of bristling hairs, spreading their wings or opening a gaping mouth. Sometimes the effect is not merely illusory but has practical results: Toads gulp air in order to make themselves too large for a snake to swallow; blowfish, puffers and porcupine fish inflate themselves into hard, spiky balls, causing predators to pass them by.

It is the pure bluffers, however, that are most impressive in their ingenious defenses. Many are harmless imitators capable of transforming themselves into semblances of frightening predators. Their performances are called Protean displays after the legendary Greek god Proteus, who kept escaping from his enemies by changing his form. Some hole-nesting birds mimic snakes by hissing loudly when disturbed. Tropical hawkmoth caterpillars when threatened reveal snakelike heads that are uncannily convincing, and they also move back and forth like vipers about to strike. One species of lantern bug has evolved a monstrous false head resembling that of a miniature alligator, complete with reptilian eyes and open jaws.

Many animals make use of flashes of color to startle or confuse an attacker, suddenly exposing bright patches or

patterns that contrast sharply with their normal camouflage. One of the most highly developed bluffing devices belongs to the moths and caterpillars that reveal frightening "eyes" when a predator comes too close. For protection, one hawkmoth that rests on trees exposes two huge false eyes on its inner wings and moves rhythmically up and down, suggesting to small birds the specter of an owl suddenly appearing from its hole in the trunk. The peacock butterfly, which usually looks like a dead leaf, when disturbed opens and closes its outer wings, displaying four blinking "eyes" instead of two—and for good measure rubs its wing edges together to produce a hissing sound. Imitation eyes, or eyespots, are not the exclusive property of insects: Numerous tropical fishes have them along their bodies or near their tails.

The eyespots that occur near the tails of many fishes and insects are designed not so much to frighten predators as to deflect bites away from the real head so that the owner can escape without harm or at least with its most vital parts intact. Some butterflies have developed the ruse in almost unbelievable detail: Their rear wingtips have the appearance of heads with false antennae and eyes; to heighten the fraud, one species quickly reverses its position on alighting and wiggles its fake antennae so that a watching bird will mistake its escape direction, allowing it to get away with a torn rear wing at most. Other techniques of deflection are practiced by snakes that raise and move their tails as if they were heads, and by lizards whose enticing tails break off in an attacker's mouth and continue to wiggle while the owner gets away.

Perhaps the ultimate deception is that of playing dead, commonly practiced by opossums and also by some insects, birds and snakes. The unexpected absence of movement often stills the killing urge in a predator—which may become indifferent, or may hesitate long enough for the bluffer to leap unexpectedly to life and escape.

Automeris io moth

Eyes on the Enemy

Eyespots—the circular patterns that appear on many insects—can cause a predator to hesitate disadvantageously or to flee in fright. To the bird that preys on moths and butterflies, the false eyes may suggest one of its own predators such as an owl or a cat. Many butterflies and moths display their eyespots suddenly, adding surprise to the stimulus of fear. However, eyespots alone do not guarantee safety: Inquisitive predators can become accustomed to them and call an insect's bluff. Consequently, many moths and butterflies have additional defenses.

The owl moth and the puss moth caterpillar have cryptic coloration as a primary defense—a predator must come close to encounter the moth's eyespots, and the caterpillar only creates its startling clown face when it is actively disturbed. The odiferous defense of the swallowtail caterpillar is a secondary system, and the hairstreak butterfly reinforces its back-to-front masquerade by wiggling its rear end to attract attention away from its head.

A hairstreak butterfly reveals false eyes, antennae and legs on its hind wingtips, a part of its anatomy that it would not be fatal for it to lose. Disconcertingly, when the insect takes off, it flies in the "wrong" direction.

Cryptically patterned, the owl moth
above adds insurance with eyespots
marked so that they resemble
the reflections of true eyes.

A pussmoth caterpillar (left) reacts to
a threat. It has retracted its head under
its body, which it inflates into a garish
"mouth" gaping under false eyes.

If eyespots on its thorax do not give a
predator pause, the swallowtail
caterpillar at right projects its
osmeterium—a writhing orange organ
that emits a foul smell.

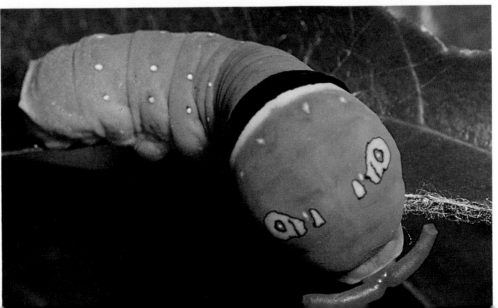

Aquatic Fireworks

Baffling predators by alternating flashes of color with crypsis is a common practice among insects. But in an unusual instance, a similar technique has evolved in the pinhead-sized copepod sapphirina. The sunlit surface waters of tropical and subtropical seas teem with these tiny crustaceans, and like all copepods—possibly the most numerous invertebrates in the world—sapphirina is a basic commodity in the oceanic food chain. To diminish its losses to predators, the species is equipped to switch from being transparent to producing dazzling displays of mini-fireworks and then revert to its usual state of transparency. The bursts of color are dependent on the direction from which light strikes minuscule platelets in dorsal skin cells of the constantly moving creature: When these platelets are illuminated from certain angles, the light is refracted and the animal begins to sparkle with shades of red and blue. The inconsistency of the copepod's image often causes a pursuer such as a fish to seek a meal elsewhere.

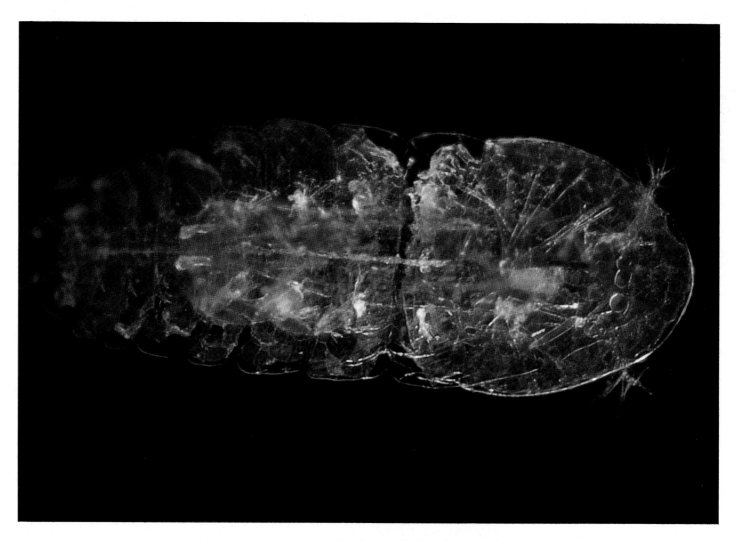

Highly magnified by a laboratory lens, a single sapphirina copepod undergoes a transformation from transparency to brilliant color. In light from one direction the copepod is transparent (above). However, when illuminated from another direction, components of cells in its back begin to deflect light rays to produce spectral colors (right, top), and finally the animal flashes like a multifaceted jewel (right, bottom).

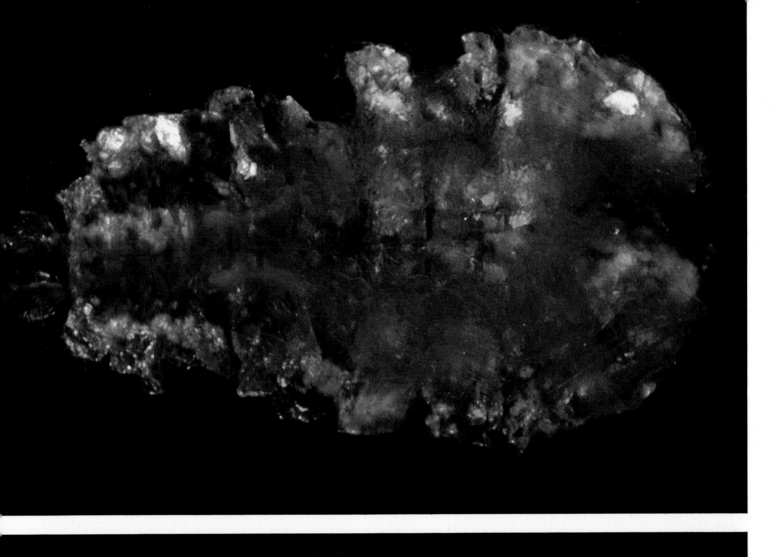

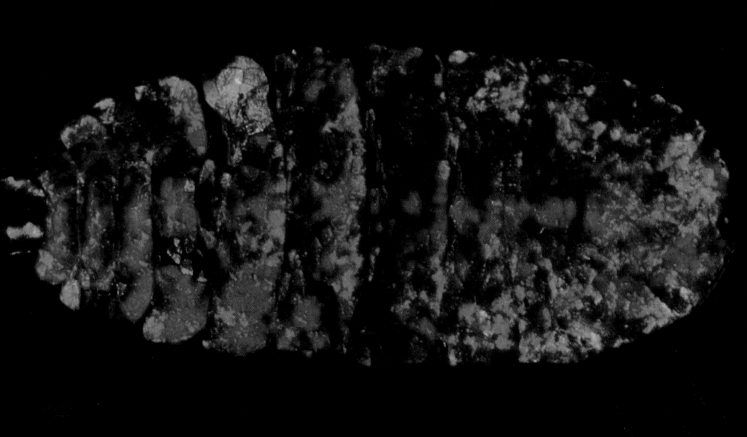

Scare Tactics

The chameleon and the mantis are experts at blending with their backgrounds and at remaining immobile to the point of invisibility. But when these two well-disguised creatures come under attack, they adopt a threatening posture, quickly exchanging their camouflage for defiant displays.

Expansion is one such effective method of intimidation. Some chameleons inflate their lungs to enlarge their entire bodies, while the sticklike mantis elevates its wings and forelegs to give an illusion of girth. Position relative to the enemy is an important part of the strategy. The blown-up chameleon often turns itself sideways so that the antagonist's view is of its bulkiest aspect, and the mantis wheels around to face its adversary with its broad forelegs raised near its open jaws as it sways to and fro. The displays of both insect and reptile are convincing enough to frighten even a human, but though the spiny forelegs of the mantis might be capable of injuring a small predator, the chameleon's act is pure bluff.

An East African mantis (above) abandons its usual low profile for a threat posture. With wings held aloft, it displays its spine-fringed forelegs with their pincers next to its bright-red mouth lining.

Jaws agape, a female Jackson's chameleon lunges and hisses like a snake. The chameleon changes color when threatened, often darkening; the sudden appearance of the contrasting color inside its mouth startles a foe.

Groups and Coalitions

The old adage that there is safety in numbers can be observed at work among animals in many striking ways. While solitary animals may be able to call on one or more defenses to outwit an enemy, other animals often gain a greater advantage by associating closely with their own kind. Living together not only ensures reproduction and eases the tasks of food gathering and nest building but also dramatically increases the odds that an individual will survive when it is threatened.

The basic grouping in the animal world is the mating pair and its young. Parents of most of the more advanced animal species form an alliance to define a home territory and defend it against all comers—rivals and predators alike. If the young are threatened, the parents may go to extraordinary lengths to protect them. A mother bear may attack savagely and unexpectedly if she suspects some danger to her cubs, and even small birds will attempt to protect their chicks by diving and pecking at an intruder or by distracting it with noisy chatter. Some birds are masters of diversion, especially of the "broken wing" ploy: When a predator approaches, one parent flaps about noisily as if injured in order to draw the attack to itself. Similarly, some fishes thrash about to feign injury when a predatory fish comes too near their young.

Many animals combine the advantages of family association with the even more effective defenses conferred by membership in still larger groups. Some fishes travel in schools; antelope, moose and other hoofed animals associate in herds; wolves and baboons live in packs; social insects like bees, wasps, ants and termites congregate in hives and mounds. A basic advantage of such mass communities is that membership lowers the odds of chance encounters with widely dispersed predators; although a group is more easily detected because of its size, its very concentration may make all but the most determined attacker think twice before acting. If a predator does take on a group, it may face a confusion of moving bodies, and often a concentration of retaliatory weapons. In a group, moreover, an animal can rely on the truism that many heads—or, more precisely, many eyes, noses and ears—are better than one. Some animals post sentinels to give the alarm; many rely instead on a mass early-warning system that consists of all members looking around or sniffing the air at random every few seconds, thus ensuring that at least one will spot a predator while it is still far enough away to allow a choice of standing ground or fleeing. When one member does detect danger, it alerts the rest with a snort, bark, special cry or, in the case of a beaver, a slap on the water with its broad tail, which sounds like a rifle shot.

Animals of different species often mingle, benefiting not only from one another's vigilance but also from complementary physical attributes. Baboons, which possess sharp eyesight as well as formidable fighting power, and antelope, which have keen senses of hearing and smell, frequently forage together in a cooperative defense against enemies. Some fairly defenseless creatures team up with species that have formidable weapons. Hermit crabs, for instance, derive protection from the poison of sea anemones that are attached to the gastropod shells in which the crabs live; the anemones in turn gain a measure of mobility.

Many small birds associate in mixed-species groups. The alarm calls of most are so alike that when one warns of an approaching hawk, all species are instantly alert—and in some mixed flocks, even dissimilar warning calls are learned and understood. Even more effective are the alarm systems of many insects and fishes, which include a release of warning chemicals that stimulate both kin and associates either to rally or to flee.

If an attacker persists, a group may adopt more active methods of defense. A school of fish may scatter in different directions, causing a predator to hesitate over which to pursue while its members get away. Conversely, many herd animals such as zebras and musk oxen cluster, forming a tight circle that prevents a lion, for example, from targeting a single victim; such a circle can be broken only at considerable risk. Small birds may distract a hawk by "mobbing"—diving in concentrated flocks at the intruder in repeated mock attacks.

Perhaps the most sophisticated group defenses are those of termites and ants. Their elaborate divisions of labor include some individuals that, using their enlarged, flat heads to plug entrance holes, act as doorkeepers, and others that, as soldiers heavily armed with powerful jaws, stings or acidic sprays, rush to defend any breach. Termite soldiers of one species wage a specialized and devastating form of chemical warfare: When they have identified approaching enemies, they eject long, liquid threads of chemicals that attract more soldiers to the scene of danger and also trap the invaders in a tangle of glue.

Hepatus crab with sea anemone

Underwater Safety

Many marine animals that are relatively defenseless band together for security or exploit the defenses of other better-armed animals. Fishes and crustaceans, for instance, may associate with jellyfish and sea anemones, both of which have impressive weapons—their tentacles contain stinging cells that poison their prey. Many such relationships between species are commensal: The protector neither benefits from nor is harmed by the animal it is protecting. The protected animal, however, must be able to neutralize the more formidable animal's defenses or it will become just another victim. Some fishes and crustaceans are immune to the poisons of the creatures with which they coexist; others must be constantly wary, staying close enough to them for defense but avoiding actual contact.

A less risky association is the grouping of a single species, such as the schooling of fish. Out of a mass of identical individuals, a predator must choose which to pursue. If the threatened fish scatter in all directions as they often do, usually only one fish is likely to be caught—a small price to pay for the survival of the rest.

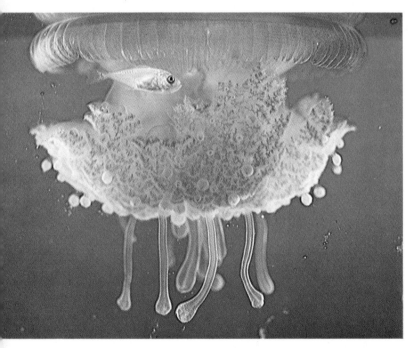

A carangid (above) hovers close to a jellyfish but keeps a healthy distance from the tentacles dangling below. The squat lobster larva at right hitches a ride on the upper side of a jellyfish—the tentacles are underneath.

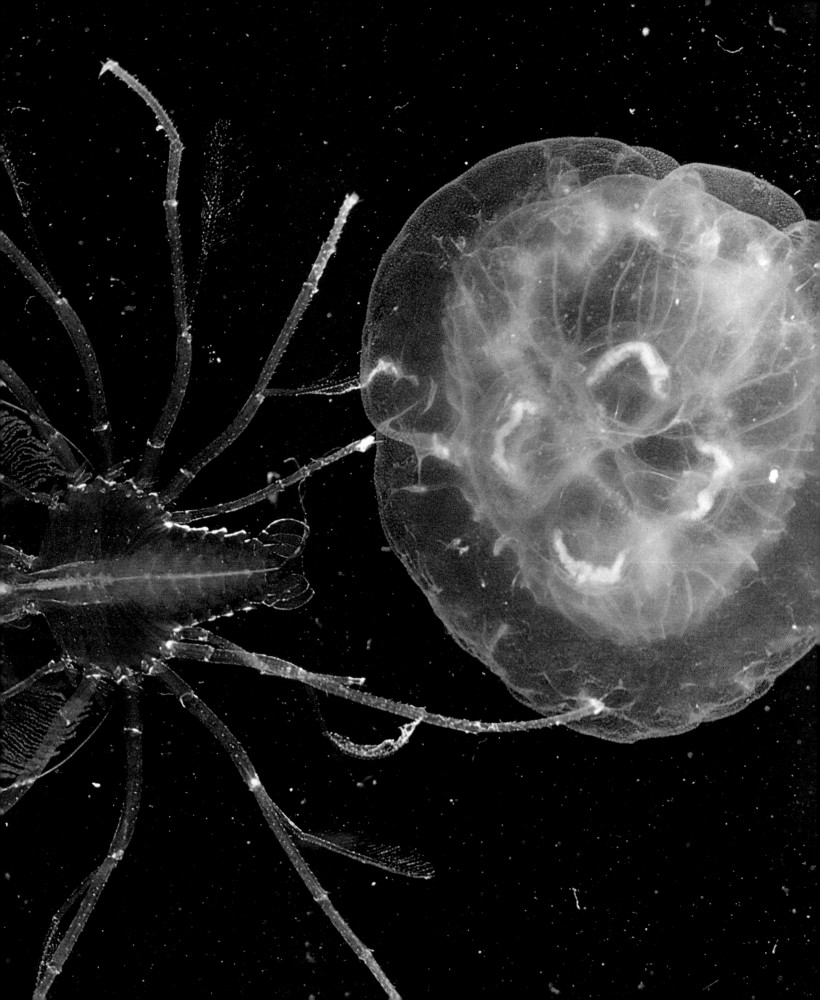

A boxing crab (above) clutches sea anemones as weapons to be used in its own defense. The crab is immune to the anemone's stinging cells and in an unknown way inhibits the anemones from devouring it.

An anemone fish swims among the tentacles of a sea anemone (above). The fish has acclimated to the anemone and has acquired a protective coating of mucus. It is now impervious to its protector's stings.

Schooling by fish (overleaf) compels a predator to choose which fish to pursue, giving the school time to escape.

The Aegis of Ants

Well-organized and efficient ant colonies offer ideal protection not only for individual ants but also for many other kinds of insects—if they can associate with the ants without being attacked. Some beetles that coexist with ants rely on their thick exoskeletons to armor them against injury, while crickets that reside in ant colonies depend on their agility to stay out of the way of their hosts' biting jaws. And some species of defenseless caterpillars offer the ants a kind of payment in exchange for their protection.

The caterpillars hatch from eggs laid on a plant that the ants frequent. When an ant discovers a larva it begins stroking a section of the caterpillar's back with its forelegs and antennae and then sips the sweet glandular secretion that is exuded. In return for this treat, the ant defends the caterpillar from predators and parasites. One species of caterpillar, however, induces hospitable worker ants to take it home to the nest. When safely installed, the caterpillar begins to feed on its host's larvae.

An ant attends the pupa of a lycaenid butterfly (above). In this stage of development the lycaenid cannot be "milked," but the ant tolerates its presence until the pupa hatches into a butterfly.

An ant gets compensation from an imperial blue caterpillar (right). It strokes a dorsal segment of the caterpillar's body, stimulating a gland to secrete a fluid that the ant relishes.

100

Fire Hazard

A single bee is rarely cause for alarm, but the stings of an antagonized swarm—especially of some African species—are evidence of the injury that the smallest animals can inflict when they act in concert. Colonies of fire ants, aggressive insects no more than a quarter inch long, have even acquired the status of a major menace.

Accidentally imported into the United States from South America, the red fire ant has flourished in the Southern states, building mounds as high as three feet that can house up to 100,000 ants. At the slightest provocation, the ants attack, first biting and then pivoting to aggravate the wound with a painful, burning sting. Their assaults have killed calves, pigs and chickens as well as allergic humans. Although valued in their native habitat as destroyers of other insects, they have few natural enemies in the United States; rapid breeders resistant to environmentally acceptable pesticides, they have recently been increasing their range at a rate of up to 30 miles per year.

Few animals dare to approach a bee colony like the one at left in a euphorbia tree in Uganda. If any member is interfered with, a group attack on the perpetrator is likely, and the resulting stings can be lethal.

A mass of fire ants continues to simmer even when afloat, constantly revolving so that all can breathe. The sting of these ants is perilous and they have become hazards in the United States.

Numerical Edge

Some species of gulls and terns, members of the same family of sea birds, nest together in large colonies—and although the masses of birds guarding their eggs and chicks are highly visible targets for predators, the gregarious brooding can have advantages. The colonial nesting of Heermann's gulls and elegant terns on isolated Raza Island in the Gulf of California probably evolved under pressure from predators such as the peregrine falcons and bald eagles that once inhabited the region. The larger and stronger gulls had enough numbers to mob an intruder and drive it off; the terns nesting in their midst could depend on the gulls to fend off an enemy before it could attack a tern egg or chick. Although the populations of falcons and eagles have greatly diminished through the harmful effects of pesticides, Heermann's gulls and elegant terns still nest together. Most of their defensive tactics, however, are now employed against each other in battles over territory.

Heermann's gulls and elegant terns (right) nest together on the island sanctuary that is a major breeding ground for the two species. The noise and activity of such sociable nesting probably stimulates and accelerates reproduction. Due to territorial battles, however, there are substantial casualties on both sides: On the edge of its colony, the tern chick above is being attacked by an encroaching gull.

Season on the Plain
by Franklin Russell

Franklin Russell, a former journalist, won acclaim for his 1961 book Watchers at the Pond, *a story of life through the four seasons around a North American pond. Russell has since written a number of other books about nature employing a storytelling technique, among them* Season on the Plain, *a study of wildlife on the East African savanna. The following selection from that book describes the response of a herd of Thomson's gazelles to a pair of hunting hyenas.*

The gazelles had come steadily south in the grip of sweeping rains, though sudden, quick spasms of drought and heat sometimes held them uncertain. Throughout the migration, the buck gazelle fought every step of the way to establish territory. He fought for territory at a water hole, even though the gazelles were not drinking—they scarcely ever needed water. He locked horns with other buck

gazelles to ensure his modest territorial demand. They pushed and shoved until one buck broke away and trotted off. Neither animal was hurt in these encounters. He fought for territory along the banks of the river flanking the woodlands and at the edge of the plains themselves, and now, deep into the plains, he was still ready to fight his colleagues for featureless pieces of earth.

This was easy enough, a simple series of decisions to make, but the decisions became complicated whenever a hunter appeared. An old female hyena had come clumping into view one midday, walking among gazelles grazing and gazelles resting. Her udders hung low, and with her head-down stance and her heavy footsteps, she was making no effort to conceal herself. The buck watched her warily but divined that she was not dangerous. He resumed grazing. The hyena passed so close that she could have rushed forward and seized him easily enough, but she disappeared, passing clear through the center of the unconcerned herd.

In the late afternoon of the same day, though, a pair of hyenas had appeared from the north, their rounded ears on the horizon announcing their approach. Instantly, the buck was alert. All the other bucks sensed his concern. Every head came up. The buck advanced a dozen hesitant steps, snorted, then held his ground. This was known danger, not just because there were two hyenas—hyenas frequently hunted alone—but because the quality of their interest was totally different. Now, the shape-up would begin, the hunters estimating the wariness of the gazelles, while the gazelles counterestimated the danger of attack. Instantaneous flight from all possible sources of danger was wasteful and time-consuming, and compared with the

number of animals killed in any series of attacks, not worth the effort. Besides, the hyenas—and most of the other great meat eaters—frequently did not attack.

Normally, the buck would have allowed the hyenas to come within five hundred feet before turning himself to flee. But something about these two hyenas—youngsters hunting together—communicated their hunger to him, and the buck prepared to react correctly. He waited to find out whether they intended to approach steadily or whether they had decided to make a rush. His response to a steady approach would be one burst of flight before he turned and faced them again to reappraise their intentions. They rushed.

Instead of fleeing, the buck began dancing. With legs stiff and neck bowed, his dancing was an odd up-and-down movement that held him in one place. The dancing, or

stotting, or pronking—there was no adequate word for it—spread instantly to a hundred other gazelles. An undulating mass of brown backs and striped sides leaping insanely up and down became the unbelievable backdrop to the attack of the two killers. The inexperienced hyenas did not understand that they had already lost the hunt. They ran at full speed, confident of a kill.

Magically, the dancing stopped in relays, and the tiny antelopes darted away one by one. Their flight, once begun, was many times faster than the top speed of the hyenas, who were too young to know that gazelles rarely ran very far and were not capable of long-distance flight. As the attack was pressed forward, it became apparent that the hunters had not made one crucial decision: they had

not chosen a specific animal to kill. Their attack became aimless as some of the nearer gazelles began pronking again. The dancing spread to involve thousands. The hyenas ran into a sea of dancing, darting, zigzagging animals which flowed together into a corporate victim without identity and, therefore, not vulnerable.

Rain clouds fled south. The hyenas became exhausted before any of the gazelles were tired, and they stood together, tongues lolling in the now-hot sun, watching as the gazelles formed up two hundred feet away, the buck at their head again. He divined that the danger was over and dropped his head and began grazing. One by one the gazelle heads went down. The hyenas seemed to understand they had been dismissed and ambled away.

A herd of wapiti (above) grazes in Wyoming. Their hooves and antlers defend them against predators, but their major threat is insufficient forage.

Deceptively docile in appearance, the African buffaloes guarding a calf below are highly dangerous animals that weigh up to a ton and are preyed upon only by lions. The bird on one buffalo's back is an oxpecker. When it spots an intruder in the distance, its cries warn the buffaloes.

Hooves and Horns

Wapiti fight cougars with hooves and antlers, and the herds can run swiftly, swim well and cross vast distances of rough territory without tiring. With calves clustered in the center and 900-pound bulls foraying outward stabbing with their horns, the musk oxen's circle can withstand the siege of a wolf pack. However, as in other once-populous North American herd animals, these defenses proved useless against the guns of men, and with their numbers diminished, musk oxen and most wapiti are now protected in preserves. At the same time, their predators are increasingly threatened species.

In Africa, however, buffaloes often still defend themselves against the predations of lions, the only animals that consistently dare to hunt them. The aggressive and alert buffaloes, with their testy dispositions and lethal horns, will charge almost anything that moves, including a stalking lion. The lions are more successful if several cats cooperate, but even then a herd zealously protecting its calves presents a dangerous bulwark and old bulls that wander alone are the most likely to be brought down.

Musk oxen (overleaf) stand in a defensive circle. To a predator attacking from any direction, the formation presents a united front.

Fight or Flight

No matter how strong or elaborate its defenses, an animal faced by a superior or even an equal adversary usually tries to get away. If that proves impossible, the animal must make a stand and retaliate as best it can. The bodies of most animals can prepare them instantly for either action: Fear or surprise steps up their heart rates, sending blood to muscles, brain and lungs so that they are ready for maximum effort of any kind. Whether an animal flees or fights depends on the nature of the attacker and the circumstances of the attack. The choice may also hinge on instinct or the animals' capacity for judgment.

In many species, particularly those that are small or otherwise vulnerable, the choice is predetermined: The animal instinctively tries to escape by whatever means of locomotion it commands—running, swimming, flying, leaping up or dropping down. Some fishes and mammals depend on sheer speed, making off in a straight line of escape. Others rely on different tactics. To get out of harm's way, many spiders and caterpillars drop on silken threads spun out of their bodies, climbing back up on their lifelines when the danger has passed. Grasshoppers and kangaroos execute spectacular evasive jumps. When approached by a predatory starfish, even such sluggish creatures as limpets and snails thrust their "feet" out of their shells to perform acrobatic tumbles and leaps.

Other animals combine flight with unexpected movements or distractions that frustrate pursuit. Moths and butterflies, whose relatively large wings and slow flying speed make them somewhat vulnerable, always fly an erratic course to confuse a pursuer. Many insects are cryptic at rest, but in flight they can display a flash of color that deceives a predator's eye by vanishing abruptly when the insect alights and becomes cryptic again. Rabbits, birds and larger animals like deer and zebras often zigzag, jump, stop, turn or dodge in and out of cover to make themselves harder to catch.

The white tail patches displayed by cottontail rabbits, white-tailed deer and some species of birds are also used as flash colors to baffle a pursuer. In addition, they may serve as alarm signals and as messages to the enemy: "You have been seen and you might as well withdraw—or prepare for a long, hard chase." Similarly, the strange, stiff-legged prancing, or stotting, of Thomson's gazelles can be used as an alarm, as a distraction and then as a lure to entice a predator away from fawns. Or it can be used as a means both of informing the potential attacker that it has been observed and of testing its intent. The message is not lost on experienced predators: Tigers and lions often ignore gazelles or other prey that have noticed their presence, preferring to conserve their energy for a target that offers a greater chance of success.

If a predator does successfully ambush, overtake or corner its prey, there is no defense left for most species but to fight. Even harmless-seeming animals can retaliate with a swiftness and strength that can stun an attacker and persuade it to retreat. Like the weakling schoolboy who finally turns on the class bully, a jackrabbit may abruptly stop and plant a painful kick on a coyote's sensitive nose; giraffes, ostriches and kangaroos can administer bone-crushing and often fatal blows with their feet. Indeed, some herd animals prefer to stick together, making a stand with their hooves and horns, rather than risk an exhausting chase during which individuals may become separated from their fellows and pulled down. Many animals that must contend with several kinds of predators are able to tailor their defenses to the size and capabilities of each. A female Thomson's gazelle with a fawn, for example, might attack a jackal to drive it away, attempt to distract a larger hyena, but run from a lion, cheetah or wild dog, relying on her greater agility and stamina.

The effectiveness of an animal's defense is not necessarily relative to its size. It is a curious fact that some of the smallest creatures can also be among the deadliest: Scorpions, black widow spiders and short-tailed shrews, which use their venom primarily to paralyze the small animals that they prey on for food, can turn their chemical weapons on much larger targets when they are threatened, sometimes with fatal results.

Among some animals, such as bees and zebras, when one is attacked, the others may go to the rescue. But most prey species ignore a victim. Predation is usually in the best interests of a group because, as zoologist George Schaller has written: "The ultimate factor is the habitat, the food supply. . . . Predators . . . hold numbers below a level at which disease and starvation become important regulating factors. . . . They help maintain an equilibrium in the prey population within the limits imposed by the environment. . . ." And often it is the sick, weak or otherwise expendable members of a species that lose the battle for survival, to the benefit of the species as a whole.

Raccoon confronting cougar

The Last Ditch

Most active combat among animals takes place between members of the same species: Groups or individuals contend for territory and, during the breeding season, males may fight each other for dominance and for possession of females. These contests seldom result in fatalities or even in serious injury to the participants. Battles between predator and prey, however, can become desperate fights to the death. The potential victim will usually attempt to flee from attack; but if the predator is almost upon the prey or if the prey has been cornered or actually seized, it may resort to aggressive retaliation. If the prey can inflict a painful or debilitating wound, its predator may abandon the attack as not being worth the trouble.

Invasion of territory by another species is also cause for aggressive defense, especially if helpless young are nearby. A badger, for instance, will defend the area of its den with tooth and claw, discouraging any curious animal from further exploration.

An inexperienced grizzly bear cub rears back from an annoyed and snarling badger. Most animals learn to avoid the bad-tempered badger's territory—its ferocious defense is respected even by humans.

A struggling water snake tries to bite its voracious captor, a snapping turtle (right), in a retaliatory attempt to defend itself. Usually the snapping turtle preys on the snake, as well as on birds and fish, when it is in the water.

White Fang

by Jack London

Impoverished and adventurous, Jack London joined the Klondike gold rush in 1897 when he was 20 years old, and began to write of his experiences in stories and novels that would make him famous. Among his most popular books are The Call of the Wild *and its sequel,* White Fang. *In the following excerpt from* White Fang, *a wolf cub that has just finished a meal of ptarmigan chicks finds out that a mother ptarmigan can be an awesome adversary.*

He encountered a feathered whirlwind. He was confused and blinded by the rush of it and the beat of angry wings. He hid his head between his paws and yelped. The blows increased. The mother ptarmigan was in a fury. Then he became angry. He rose up, snarling, striking out with his paws. He sank his tiny teeth into one of the wings and pulled and tugged sturdily. The ptarmigan struggled against him, showering blows upon him with her free wing. It was his first battle. He was elated. He forgot all about the unknown. He no longer was afraid of anything. He was fighting, tearing at a live thing that was striking at him. Also, this live thing was meat. The lust to kill was on him. He had just destroyed little live things. He would now destroy a big live thing. He was too busy and happy to know that he was happy. He was thrilling and exulting in ways new to him and greater to him than any he had known before.

He held on to the wing and growled between his tight-clenched teeth. The ptarmigan dragged him out of the bush. When she turned and tried to drag him back into the bush's shelter, he pulled her away from it and on into the open. And all the time she was making outcry and striking with her wing, while feathers were flying like snowfall. The pitch to which he was aroused was tremendous. All the fighting blood of his breed was up in him and surging through him. This was living, though he did not know it. He was realizing his own meaning in the world; he was

doing that for which he was made—killing meat and battling to kill it. He was justifying his existence, than which life can do no greater; for life achieves its summit when it does to the uttermost that which it was equipped to do.

After a time, the ptarmigan ceased her struggling. He still held her by the wing, and they lay on the ground and looked at each other. He tried to growl threateningly, ferociously. She pecked on his nose, which by now, what of previous adventures, was sore. He winced but held on. She pecked him again and again. From wincing he went to whimpering. He tried to back away from her, oblivious of the fact that by his hold on her he dragged her after him. A rain of pecks fell on his ill-used nose. The flood of fight ebbed down in him, and, releasing his prey, he turned tail and scampered off across the open in inglorious retreat.

Speed and Strategy

Many animals, especially herd animals, do not run away at the first sight of a predator. If the enemy is far enough away, it may be ignored; if it approaches more closely, the prey may freeze and watch it; and finally, when the predator reaches what has been called the "flight threshold," the prey will flee until it has reestablished a safe distance between itself and the predator. Newborn zebras, wildebeest and other prey animals have no intuitive fear of predators, and they must learn—through experience—the different flight threshold for each enemy.

Herd animals take other precautions to avoid an exhausting chase. They shun dense cover where a predator might be lurking and may also stay clear of water holes at dusk, the time of day when many predators begin to hunt. If flight does become necessary, speed is not always the only key to escape. By greater endurance an animal may outlast a predator. In addition, some animals confuse and tire a pursuer by running and leaping in an erratic pattern. Gazelles have even been known to evade the final rush of an attacking lion by jumping over the startled cat.

A pack of hyenas circle a lone zebra in a concerted attack (above). The besieged zebra will fight back—kicking with front and hind feet.

A bighorn sheep sprints away from a cougar. The bighorn's escape often depends on its ability to race across rocks and leap over chasms. Its acute senses usually warn it of a predator ahead of time.

120

Ambushed by a lioness, warthogs quickly gear up to dust-raising speed (below). Warthogs sometimes charge an attacker with their sharp canines.

With seemingly effortless leaps, two Grant's gazelles (overleaf) try to escape from a pursuer. The easy grace of their flight belies their fear.

Wild Animals at Home

by Ernest Thompson Seton

Accounts of North American wildlife by naturalist Ernest Thompson Seton have been a popular introduction to the world of nature for generations of young readers. In 1897 Seton visited the Badlands of North Dakota, where he witnessed the encounter described below between greyhounds and a blacktail, or mule deer, and her fawns. In this episode from Wild Animals at Home, *the speed of the greyhounds, which cannot be restrained from the chase, is matched against the surefootedness of the deer.*

Away went the Blacktail, bounding, bounding at that famous beautiful, birdlike, soaring pace, mother and young tapping the ground and sailing to land, and tap and

sail again. And away went the greyhounds, low coursing, outstretched, bounding like bolts from a crossbow, curving but little and dropping only to be shot again. They were straining hard; the Blacktail seemed to be going more easily, far more beautifully. But alas! they were losing time. The greyhounds were closing; in vain we yelled at them. We spurred our horses, hoping to cut them off, hoping to stop the ugly, lawless tragedy. But the greyhounds were frantic now. The distance between Bran and the hindmost fawn was not forty feet. Then Eaton drew his revolver and fired shots over the greyhounds' heads, hoping to scare them into submission, but they seemed to draw fresh stimulus from each report, and yelped and bounded faster. A little more and the end would be. Then we saw a touching sight. The hindmost fawn let out a feeble bleat of distress, and the mother heeding dropped back between. It looked like choosing death, for now she had not twenty feet of lead. I wanted Eaton to use his gun on the foremost hound, when something unexpected happened. The flat was crossed, the Blacktail reached a great high butte, and tapping with their toes they soared some fifteen feet and tapped again; and tapped and tapped and soared, and so they went like hawks that are bounding in the air, and the greyhounds, peerless on the plain, were helpless on the butte. Yes! rush they might and did, and bounded and clomb, but theirs was not the way of the hills. In twenty heartbeats they were left behind. The Blacktail mother with her twins kept on and soared and lightly soared till lost to view, and all were safely hidden in their native hills.

Credits

Cover—R. Hermes, Photo Researchers, Inc. 1—M. Fogden, Bruce Coleman, Inc. 5, 9—P. Ward, B.C., Inc. 14 (top)—Oxford Scientific Films; (bottom)—Oxford Scientific Films, B.C., Inc. 17—P. Ward, B.C., Inc. 18 (left)—C.J. Cole; (right) W. Ruth, B.C., Inc. 19 (top)—T. McHugh for the Steinhart Aquarium, P.R., Inc.; (bottom)—Oxford Scientific Films. 20—P. Chesher, P.R., Inc. 20–21—D. Lyons, B.C., Inc. 22—Oxford Scientific Films, B.C., Inc. 23 (top)—H. van Lawick, B.C., Inc.; (bottom)—Oxford Scientific Films. 24–25—Oxford Scientific Films. 26—L. Stone, Animals Animals. 26–27—M. P. Khal, B.C., Inc. 28—J. Burton, B.C., Inc. 28–29—K. Tweedy-Holmes, Animals Animals. 33—R. Dunne, B.C., Inc. 34—P. Ward, B.C., Inc. 35 (top)—Oxford Scientific Films; (bottom) R. Sterling, P.R., Inc. 36–37—M. Fogden, B.C., Inc. 38—Oxford Scientific Films. 38–39—M. Fogden, B.C., Inc. 40–41—C. Mann, P.R., Inc. 41—P. Ward, B.C., Inc. 42–43—Oxford Scientific Films. 43 (top left)—P. Ward, B.C., Inc.; (top right) Oxford Scientific Films. 44—A. Power, B.C., Inc. 44–45—T. McHugh, Steinhart Aquarium, P.R., Inc. 46 (bottom left)—C. Roessler, Animals Animals; (bottom right)—B. Brander, P.R., Inc. 46–47—Oxford Scientific Films. 47 (bottom)—R. Borland, B.C., Inc. 48–51 (top)—Oxford Scientific Films. 50–51—R. Mendez, Animals Animals. 52–53—L. Battaglia, P.R., Inc. 54–55—Stouffer Productions, Animals Animals. 55 (top)—G. Jones, B.C., Inc.; (bottom)—F. Whitehead, Animals Animals. 60—T. Brakefield, Animals Animals. 60–61—Peter B. Kaplan. 63—J. Bockowski, Animals Animals. 64 (top)—M. Fogden, B.C., Inc.; (bottom)—P. Ward, B.C., Inc. 65—M. Fogden, B.C., Inc. 68–69—P. Ward, B.C., Inc. 70 (left)—M. Fogden, B.C., Inc.; (top right)—P. Ward, B.C., Inc.; (bottom right)—R. Mendez, Animals Animals. 71 (top left)—J. Robinson, P.R., Inc.; (top right)—P. Ward, B.C., Inc.; (center left)—P. Ward, B.C., Inc.; (center right)—R. Mendez, Animals Animals; (bottom)—T. Taylor, P.R., Inc. 72 (top)—A. Blank, B.C., Inc.; (bottom)—M. Fogden, B.C., Inc. 73—Z. Leszczynski, Animals Animals. 74–75—Stouffer Productions, Animals Animals. 75—J. & D. Bartlett, B.C., Inc. 78—T. McHugh, Steinhart Aquarium, P.R., Inc. 79 (top)—J. Robinson, P.R., Inc.; (bottom)—P.R., Inc. 80 (top)—A. Dignan, B.C., Inc.; (bottom)—J. Himmelstein, B.C., Inc. 80–81—J. Markham, B.C., Inc. 82–83—Oxford Scientific Films. 84–85—J. Robinson, P.R., Inc. 86 (left)—M. Fogden, B.C., Inc.; (bottom right)—K. Preston-Mafham, Animals Animals. 86–87—S. Bisserot, B.C., Inc. 87 (bottom)—M. Fogden, B.C., Inc. 88–89—Oxford Scientific Films. 90—P. Ward, B.C., Inc. 91—Z. Leszczynski, Animals Animals. 93—R. Mariscal, B.C., Inc. 94–95—Oxford Scientific Films. 96—J. Burton, B.C., Inc. 97—R. Mariscal, B.C., Inc. 98–99—S. Summerhays, P.R., Inc. 100–101—Oxford Scientific Films. 102—S. Trevor, B.C., Inc. 103—C. Lockwood, Animals Animals. 104–105—Thase Daniel. 110–111 (top)—M. Newman, Animals Animals; (bottom)—J. & D. Bartlett, B.C., Inc. 112–113—F. Bruemmer. 115, 116—Stouffer Productions, Animals Animals. 117—Tom Brakefield. 120—Stouffer Productions, Animals Animals. 121 (top)—S. Fresco, B.C., Inc.; (bottom)—G. Schaller, B.C., Inc. 122–123—J. van Wormer, B.C., Inc.

The photographs on the endpapers are used courtesy of Time-Life Picture Agency and Russ Kinne and Stephen Dalton of Photo Researchers, Inc., and Nina Leen.

The film sequence on page 8 is from "Garden Jungle," a program in the Time-Life Television series *Wild, Wild World of Animals*.

ILLUSTRATION on pages 6–7 is by Rockwell Kent and Abbott H. Thayer. The illustration on page 10 is by Abbott H. Thayer, those on pages 12–13 are by Mark Kseniak, that on page 31 is by Charles B. Slackman. The drawing on page 57 is by Rudyard Kipling. The illustrations on pages 67 and 76–77 are by Ray Cruz, those on pages 106–109 and 118–119 are by John Groth, that on pages 124–125 is by André Durenceau.

Bibliography

NOTE: An asterisk at the left means that a paperback volume in also available.

Caras, Roger, *North American Mammals*. Meredith Press, 1967.

———, ed., *Protective Coloration and Mimicry*. Barre-Westover, 1972.

Cochran, Doris M., *Living Amphibians of the World*. Doubleday, 1961.

Cott, Hugh B., *Adaptive Coloration in Animals*. Methuen, 1957.

Droscher, Vitus B., *The Mysterious Senses of Animals*. Dutton, 1965.

Edmunds, M., *Defence in Animals*. Longman Group Limited, 1974.

Farb, Peter and the Editors of Time-Life Books, *The Land and Wildlife of North America*. Time-Life Books, 1966.

Fogden, Michael and Patricia, *Animals and their Colors*. Crown, 1974.

Goetsch, Wilhelm, *The Ants*. University of Michigan Press, 1957.

Hamilton, W. J., *American Mammals*. McGraw-Hill, 1939.

Klots, Alexander B. and Elsie B., *Living Insects of the World*. Doubleday, 1959.

Linsenmaier, Walter, *Insects of the World*. McGraw-Hill, 1972.

*Lorenz, Konrad, *On Aggression*. Harcourt, Brace, Jovanovich, 1966.

*Matthiessen, Peter, *The Tree Where Man Was Born*. Dutton, 1974.

Palmer, Ralph S., *The Mammal Guide*. Doubleday, 1954.

*Porter, Eliot, *The African Experience*. Dutton, 1974.

*Portmann, Adolf, *Animal Camouflage*. University of Michigan Press, 1959.

Rowland-Entwistle, Theodore, *The World You Never See: Insect Life*. Hamlyn, 1976.

Schaller, George B., *Serengeti: A Kingdom of Predators*. Knopf, 1972.

Schwarzkopf, Chet, *Heart of the Wild*. Caxton, 1969.

Shuttlesworth, Dorothy E., *Animal Camouflage*. Natural History, 1966.

Street, Philip, *Animal Weapons*. Taplinger, 1971.

Teale, Edwin Way, *Strange Lives of Familiar Insects*. Dodd, Mead, 1962.

von Frisch, Otto, *Animal Camouflage*. Watts, 1973.

Walker, Ernest P., *Mammals of the World*. Johns Hopkins University Press, 1975.

Wicker, Wolfgang, *Mimicry in Plants and Animals*. McGraw-Hill, 1968.

Index